Penguin Education
Penguin Library of Physical Sciences

Nuclear Reactions
W. M. Gibson

Advisory Editor
V. S. Griffiths

General Editors
Physics: N. Feather
Physical Chemistry: W. H. Lee
Inorganic Chemistry: A. K. Holliday
Organic Chemistry: G. H. Williams

Nuclear Reactions

W. M. Gibson

Penguin Books

Penguin Books Ltd, Harmondsworth,
Middlesex, England
Penguin Books Inc., 7110 Ambassador Road,
Baltimore, Md 21207, USA
Penguin Books Australia Ltd,
Ringwood, Victoria, Australia

First published 1971
Copyright © W. M. Gibson, 1971

Filmset in Monophoto Times by Photoprint Plates Ltd,
Rayleigh, Essex
Made and printed in Great Britain by Butler and Tanner Ltd,
Frome and London

Contents

Editorial Foreword

For many years, now, the teaching of physics at the first-degree level has posed a problem of organization and selection of material of ever-increasing difficulty. From the teacher's point of view, to pay scant attention to the groundwork is patently to court disaster; from the student's, to be denied the excitement of a journey to the frontiers of knowledge is to be denied his birthright. The remedy is not easy to come by. Certainly, the physics section of the Penguin Library of Physical Sciences does not claim to provide any ready-made solution of the problem. What it is designed to do, instead, is to bring together a collection of compact texts, written by teachers of wide experience, around which under-graduate courses of a 'modern', even of an adventurous, character may be built.

The texts are organized generally at three levels of treatment, corresponding to the three years of an honours curriculum, but there is nothing sacrosanct in this classification. Very probably, most teachers will regard all the first-year topics as obligatory in any course, but, in respect of the others, many patterns of interweaving may commend themselves, and prove equally valid in practice. The list of projected third-year titles is necessarily the longest of the three, and the invitation to discriminating choice is wider, but even here care has been taken to avoid, as far as possible, the post-graduate monograph. The series as a whole (some five first-year, six second-year and fourteen third-year titles) is directed primarily to the undergraduate; it is designed to help the teacher to resist the temptation to overload his course, either with the outmoded legacies of the nineteenth century, or with the more speculative digressions of the twentieth. It is expository, only: it does not attempt to provide either student or teacher with exercises for his tutorial classes, or with mass-produced questions for examinations. Important as this provision may be, responsibility for it must surely lie ultimately with the teacher: he alone knows the precise needs of his students – as they change from year to year.

Within the broad framework of the series, individual authors have rightly regarded themselves as free to adopt a personal approach to the choice and presentation of subject matter. To impose a rigid conformity on a writer is to dull the impact of the written word. This general licence has been extended even to the matter of units. There is much to be said, in theory, in favour of a single system of units of measurement – and it has not been overlooked that national policy in advanced countries is moving rapidly towards uniformity under the

Système International (SI units) – but fluency in the use of many systems is not to be despised: indeed, its acquisition may further, rather than retard, the physicist's education.

A general editor's foreword, almost by definition, is first written when the series for which he is responsible is more nearly complete in his imagination (or the publisher's) than as a row of books on his bookshelf. As these words are penned, that is the nature of the relevant situation: hope has inspired the present tense, in what has just been written, when the future would have been the more realistic. Optimism is one attitude that a general editor must never disown!

N. F.

Preface

This book is intended to give, at a level appropriate to a student in the third year of an honours degree course, a general account of nuclear reactions and scattering processes. Some parts of it will be too elementary for the able student, while others may demand considerable effort from him and cause consternation to the non-mathematician. By laying out the material in a sequence in which description alternates with quantum mechanics, the author has tried to provide something for all, except possibly the most advanced mathematical physicists, who will feel a quite different sort of consternation at finding their subject so crudely summarized.

Some details in the choice of subject matter might appear arbitrary. It is therefore appropriate to mention that this book, while complete in itself, is intended to form part of a larger whole along with two companion volumes, *The Atomic Nucleus* by J. M. Reid and *Elementary Particles* by I. S. Hughes. Topics on the boundaries between the main themes of the three volumes have been shared out in a way which gives the best over-all coverage. For example, on the borders of their main subjects, *The Atomic Nucleus* covers alpha and beta decay, and the interaction of charged particles with matter (the basis of particle-detection techniques), while the present volume covers accelerating machines, and the spin dependence of nuclear forces. On the other borderline, *Elementary Particles* covers the meson theory of nuclear forces and conservation of parity, while *Nuclear Reactions* provides an introduction to isospin and charge independence. It is hoped that these arrangements have led to no serious inconveniences or obscurities.

The author's thanks are due to all those colleagues and teachers, too numerous to name, from whom he has learnt about nuclear physics, to Dr J. G. McEwen and Dr B. R. Pollard who have read the manuscript of this book, to Mrs P. Barton and Mrs L. Denning who have typed it, and most of all to his wife who has supported him in the whole enterprise.

Chapter 1
Introduction

1.1 Nuclear physics as a part of physics

The terms 'nuclear physics' and 'nuclear physicist' are capable of being interpreted in a wide variety of ways, for the boundaries of the terms are unclear. At the one end we may ask, when nuclear magnetic resonance becomes an integral part of an investigation in solid-state physics, at what point does the investigator become, or cease to be, a nuclear physicist? At the other end, at how high an energy does a proton accelerator cease to be an instrument for nuclear physics and become one for subnuclear physics or particle physics? Fortunately such questions need not be taken too seriously; how they are resolved for a particular purpose is usually a matter of merely arriving at an administratively convenient definition. So far as personal preference is concerned, the author would rather be called a physicist than a member of any of the subspecies.

When the nucleus is treated as a massive point charge of magnitude Ze at the centre of an atom, this is still atomic physics. But when properties of the nucleus other than its positive charge begin to make themselves felt, we are entering the subject of nuclear physics in its broad sense. For example, when the spin of a nucleus is observed to cause hyperfine structure in atomic spectra, this is either a nuclear effect in atomic physics or a spectroscopic technique for the study of a particular nuclear property, whichever way one chooses to look at it.

Similarly as we approach the physics of elementary particles, sometimes called subnuclear physics, nuclear physics may be said to stop when the properties of individual types of nucleus cease to matter, or alternatively it may be thought of as covering the whole field, including the study of the fundamental interactions at high energies.

In addition to these two, nuclear physics now has a third and a fourth frontier. These are closely correlated and concern its boundaries with engineering and biophysics, both of which have become growing points as a result of the discovery of nuclear fission.

So far as this book is concerned, the spectroscopic effects of nuclear spin and statistics are a part of nuclear physics, as are the interactions of protons and neutrons over a wide range of energies. But mesons, strange particles and nucleon resonances, with their symmetry schemes and conservation laws, are left for treatment in a companion volume (Hughes, 1971). In this connexion it should be pointed out that our discussion of the charge independence of nuclear forces introduces the reader to the conservation of isospin, from which grows the SU(3) symmetry scheme for elementary particles.

The boundary with biophysics is hardly touched, but that with engineering is subject to a considerable overlap in our discussion of nuclear reactors; however, this discussion is largely restricted to physical principles, and the excursions towards nuclear engineering are intended to provide a qualitative basis for possible quantitative study of the engineering aspects, which are now becoming so highly specialized.

Within the shared territories on the boundaries with other subjects there lies the whole field of techniques for observing the properties and interactions of nuclei, and the interpretation of these observations in quantum-mechanical terms – how to get at nuclei, how to observe what they do, what is observed and how to interpret it.

This field may be divided up into sections for convenience, though over-lapping of the sections is inevitable, as is some measure of arbitrariness. For example we omit from the present volume any description of the techniques for detecting the products of nuclear reactions, though we have to refer to them from time to time, and point out the existence of processes which make possible particular techniques of detection.

The technique of accelerating particles to high energies to bring about nuclear reactions is, on the other hand, covered in moderate detail with, for completeness, a description of how the technique is extended to very high energies which are of subnuclear rather than nuclear interest.

1.2   Nuclear dynamics

In describing the properties and behaviour of nuclei, one finds a natural separation of the subject into nuclear statics, dealing with the properties of nuclei as they exist in steady or nearly steady states, and nuclear dynamics, which includes processes in which nuclei are transformed from state to state, either spontaneously or as a result of a collision. The latter includes nuclear reactions, elastic and inelastic scattering, and alpha and beta decay.

However, the spontaneous processes of alpha and beta decay are strongly associated with the static properties of the decaying nuclei, and it is often convenient to consider them along with nuclear statics. This scheme has been adopted by Reid (1972), who covers the static properties of nuclei, decay processes and the shell model of the nucleus.

Again, elastic and inelastic scattering may be considered as special types of nuclear reaction, in which the products are identical with the initial particles, one of them having been raised to an excited state in the inelastic case. It turns out that the quantum-mechanical descriptions of reactions and scattering processes go well together. For that reason the title of the present volume has been used as short-hand for 'nuclear reactions including scattering processes', which means much the same as 'nuclear dynamics without alpha or beta decay'.

In accordance with this scheme, the introductory chapters of this book are followed by three chapters on the general theory of reactions and scattering processes. Following these come three more descriptive chapters on neutrons and nuclear energy sources, and finally two chapters of more academic discussion of nuclear forces and the effects of nuclear spin.

1.3 Types of interaction

The sub-division of nuclear physics according to type of process – reaction, scattering, alpha decay and so on – is only one way of subdividing it. Cutting right across this classification is an alternative description of the subject in terms of different types of interaction, a word which has special significance for the nuclear and subnuclear physicist: it implies the existence of a force or potential which is the same whether it leads to some sort of occurrence or simply to the maintenance of the *status quo*. In its everyday sense an interaction is the event in which two entities come together to make something happen. But in this specialized nuclear sense, an interaction is a field, the existence of which may in certain circumstances lead to a transition, in others to a steady state.

The three types of interaction which govern nuclear phenomena are:

(a) *The strong interaction.* This is responsible for the binding forces between nucleons in nuclei, which are discussed explicitly in Chapter 11 and shown to be charge independent. The existence of stable nuclei with mass number $A > 1$ is a static manifestation of the strong interaction, while neutron–proton scattering provides an example of the strong interaction leading to a dynamic process. The strength of the strong interaction is indicated by the value of its coupling constant $g^2/\hbar c$, which is of the order one. The constant g is analogous to charge in the Yukawa potential

$$V = g \frac{e^{-kr}}{r}.$$

(b) *The electromagnetic or Coulomb interaction.* The electromagnetic interaction leads to forces and energies somewhat smaller than those of the charge-independent strong interaction. The order of magnitude of the difference is given by the electromagnetic coupling constant $\alpha = e^2/\hbar c$, which is the Coulomb energy of two particles of charge e, separated by a distance equal to their Compton wavelength $\lambda = \hbar/mc$, expressed as a fraction of the rest energy mc^2. The constant α is also known as the fine-structure constant and has a value close to $\frac{1}{137}$. The electromagnetic interaction is responsible for the Coulomb energy of nuclei, which increases with the atomic number Z and leads to the lower over-all stability of nuclear matter in the largest nuclei, as compared with that in medium-sized nuclei. Dynamically, the electromagnetic interaction can show itself as the mechanism for reactions of photodisintegration, for gamma decay and for scattering of electrons.

(c) *The weak interaction.* This is the interaction responsible for coupling between electrons and neutrinos on the one hand, and neutrons, protons and nuclei on the other. For completeness, we put muons with the electrons and neutrinos and their antiparticles, to make the family of leptons (or light particles), and we put pions and other mesons with the neutrons, protons and nuclei to make the whole family of hadrons (which means strongly interacting particles). These two families affect each other only through the weak interaction (electromagnetic interaction between charged members being neglected), and this coupling gives rise to the phenomena of beta decay, positron decay and the related process of electron capture. The weakness of the weak interaction is indicated by the value of the coupling constant, which is of order 5×10^{-14}. This means that the energies of interaction are very small, as are the probabilities per unit time for transitions like beta decay which go through the weak interaction. Hence the time scales for beta decay are very long in comparison with the periods of order 10^{-20} second which are characteristic of nuclear processes mediated by the strong interaction.

(d) *The gravitational interaction.* Not relevant to nuclear physics, but sometimes listed as the fourth and weakest interaction in a universally valid list, is the gravitational interaction. The coupling constant for this interaction may be given in terms of the gravitational constant G, as $Gm^2/\hbar c$, which is the gravitational energy of two protons of mass m separated by their Compton wavelength, expressed as a fraction of the rest energy mc^2. Its value is extremely small, of order 10^{-39}.

We may sum up our discussion of nuclear interactions by saying that the subject matter of this book could have been described as 'the dynamical aspects of strong and electromagnetic interactions at low energy, omitting details of alpha decay and radiative transitions'.

1.4 Units

In the choice of units, the nuclear physicist is in a very flexible position, having freedom to choose the most suitable system for the particular problem in hand. This freedom, however, forces upon him a corresponding obligation to accept the choices made by those from whom he has to learn.

For example, an experimenter may have to read theoretical papers using the system in which Planck's constant and the velocity of light are given unit values. He may even find it convenient to discuss particle masses in terms of the reciprocals of their Compton wavelengths, after having grown accustomed to quoting them in units of energy. Certainly he will be plagued by the distinction between Planck's constant h and Dirac's constant $\hbar = h/2\pi$.

However, in the present text we shall retain all the ordinary dimensional constants, so that quantities appear with their usual dimensions. The one exception is mass, which is quoted either in atomic mass units or in terms of the equivalent in millions of electron volts.

The topical question of whether to use SI units has been met by adopting the most liberal interpretation of the SI recommendations, and occasionally going a little beyond it in the interests of simplicity and compatibility with the main literature of nuclear physics. Thus the centimetre is used when there is something significant to be gained by doing so, but otherwise lengths are given in metres, millimetres or femtometres (1 fm $= 10^{-15}$ m, a derived unit which started life as the fermi, defined as 10^{-13} cm and abbreviated to the same two letters). For cross-sections we have used the barn, which is allowed in conjunction with SI units as 10^{-28} m^2, but we have pointed out its more familiar definition (10^{-24} cm^2). Volumes and densities are troublesome: for some of the subjects we have discussed, there is no practical alternative to cm^3 and g cm^{-3}. Energies are almost all given in multiples of the electronvolt, which also comes into the 'allowed' category. Occasionally a magnetic field has been quoted in kilogauss, in the belief that its significance is likely to be grasped more quickly than that of the corresponding figure in teslas.

If these decisions have encouraged a little flexibility of outlook in c.g.s.-oriented readers as well as the SI-oriented, while causing nobody any unnecessary intellectual burden in the study of nuclear physics, they have been well made.

17   Units

Chapter 2
Particle Accelerators

2.1 Review of the needs and possibilities

The experimental study of nuclear physics may be said to have started with the observation of radiations from naturally occurring radioactive materials. With the identification of these radiations, and of the nuclear species from which they were emitted, a foundation was laid for the further exploration of the properties of nuclei.

This exploration proceeded through observing the interaction of moving nuclear particles with stationary nuclei, the moving particles for the first such observations being the α-particles emitted from radioactive sources, moving with velocities resulting from the process of emission. The interaction studied first was elastic scattering, which was used by Rutherford and his co-workers to establish the basic properties of nuclei, that they contain virtually all the mass of atoms and have positive charge indicated by the atomic number of the particular element.

The next step after the observation of elastic scattering was to use the same type of moving particle to induce nuclear reactions in the target nuclei. The first successful observation of such a reaction was reported in 1919 by Rutherford, who used α-particles from a radium active deposit to bring about the reaction

$$\,^4_2\text{He} + \,^{14}_7\text{N} \longrightarrow \,^{17}_8\text{O} + \,^1_1\text{H} \quad \text{(see section 4.2.1).}$$

Naturally occurring α-particles have severe limitations, however, as tools for the study of nuclear reactions, and an important step towards escaping from these limitations was taken by Cockcroft and Walton, who used artificially accelerated protons to cause the reaction

$$\,^1_1\text{H} + \,^7_3\text{Li} \longrightarrow \,^4_2\text{He} + \,^4_2\text{He} \quad \text{(see section 4.2.2).}$$

Accelerators of the Cockcroft–Walton type have continued to be used, in improved form, for experiments of increasing sophistication. The principle, which is well-known, involves using rectifiers and condensers in voltage-doubling circuits to produce d.c. voltages of up to one or two megavolts. This voltage is then applied directly to a tube along which ions can be accelerated through a single passage across the applied potential difference.

Cockcroft's and Walton's protons had kinetic energy up to only 700 keV, much less than that of the α-particles used by Rutherford. However, the protons are singly charged and face a lower potential barrier (see Chapter 10) than the doubly charged α-particles; even with 125 keV kinetic energy they were observed to have significant (if small) probability of approaching a target nucleus closely enough to cause a nuclear reaction. But the important point about using artificially accelerated particles was that the technique could be extended beyond the narrow limit imposed by a single natural phenomenon. Important extensions proved to be toward higher beam intensities for the study of processes with smaller cross-sections, and more significantly towards higher energies allowing access to new nuclear and subnuclear phenomena.

For the study of nuclear energy levels a newer type of d.c. accelerator has in part displaced the Cockcroft–Walton system. This is the van de Graaff accelerator, in which a potential difference is built up by charge sprayed onto a moving flexible belt, which carries it to the high-voltage end of an accelerating tube. This type of generator, running in nitrogen at several atmospheres pressure, can be used to produce potential differences of order 5 MV, and thus to accelerate protons or deuterons to kinetic energies of order 5 MeV. Pairs of van de Graaff generators, running in tandem, have been used to give total potential differences of up to 15 MeV. An attractive feature of the van de Graaff system, valuable in precise experiments, is that the accelerating voltage can be stabilized accurately by a beam of electrons directed from the target region towards the high-voltage terminal.

For energies upward of about 10 MeV, the various forms of resonance accelerator begin to have advantages. The idea is to obtain the required total energy by passing the particle many times through a relatively small potential difference. This potential difference may be obtained from an alternating voltage synchronized to be in the required sense when the particles are crossing between two electrodes. While the particles are inside one electrode, their kinetic energies are unaffected by the changing potential of the electrode. Thus the cumulative effect of a large d.c. voltage is obtained without the large voltage having to exist anywhere.

In the linear accelerator, the particles move along a straight path and the electrodes are cylinders with lengths tailored in proportion to the increasing velocity of the particles being accelerated. All the electrodes can then be connected to a single radio-frequency generator; alternate electrodes may be connected to opposite terminals of the generator or, for higher frequencies and higher energies, the electrodes may be operated as resonant cavities fed by interlinked generators in such a way that the particles make their whole journey riding on an accelerating travelling wave.

But the best-known resonance accelerator is the cyclotron, in which a magnetic field is used to guide the particles in circular paths so that they cross the same gap many times. The angular velocity of rotation is Be/mc, independent of the radius, so an alternating voltage of constant frequency can be used to provide the accelerating potential difference between D-shaped electrodes. The cyclotron and the machines developed from it are discussed further in the sections which follow. Since the full formal theory is given by other authors (e.g. Livingston and Blewett, 1962; Farago, 1970, especially Chapter 4), we shall emphasize those physical principles and practical considerations that are important for the nuclear physicist using an accelerator, without going into mathematical detail.

2.2 The principles of cyclic accelerators

2.2.1 *Radius of orbit*

A particle with charge e and momentum p, in a region of uniform magnetic field B perpendicular to the momentum, traces out a circular orbit of radius

$$R = \frac{pc}{Be}.$$ **2.1**

This relation, which is valid for all masses and velocities, whether relativistic or not, provides the basis for any discussion of cyclic accelerators.

While the ideal orbit for particles of a particular momentum is a circle of radius given by equation **2.1**, the actual paths followed by individual particles will differ from the ideal. The nature and magnitude of these differences is important in determining whether the orbits are stable, and whether they can be accommodated in the space available. We shall therefore devote some time now to discussing the general conditions for stability of nearly circular orbits.

2.2.2 *Vertical focusing*

In practice no magnetic field is perfectly uniform, but it is possible to make a magnet with a field having nearly perfect circular symmetry. By this we mean that the field is a function of r, the radial distance from the z-axis, and of z, the vertical coordinate along this axis, but it is independent of the azimuthal angle. Even though the main component is the axial field B_z, there will in general be a radial component B_r also. Since the magnetic field between the poles of the magnet must have zero divergence and zero curl everywhere†, the z-dependence and the radial dependence of the two components are correlated. In particular, curl **B** gives

$$\frac{\partial B_z}{\partial r} = \frac{\partial B_r}{\partial z}.$$

This is a formal way of saying that if the field has a radial dependence, the field lines must be curved. If the magnitude of B_z increases with r, the curvature is as shown in Figure 1(a), but if it decreases, the curvature is as shown in Figure 1(b).

---

†We assume that the current density of the circulating particles gives negligible curl **B**.

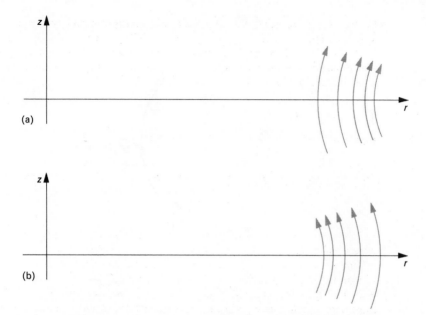

Figure 1 Curvature of magnetic field associated with radial dependence of B_z: (a) $n>0$; (b) $n<0$

The radial dependence of B_z may be specified by the field index n, which is defined by

$$n = \frac{r}{B_z}\frac{\partial B_z}{\partial r} = \frac{\partial(\log B_z)}{\partial(\log r)}.$$

It will be seen that with n so defined, B_z is proportional to r^n locally, but not necessarily over a wide range of r since n may be a function of r.

In a magnet of this type, the ideal orbit is circular, with centre on the z-axis, and in a plane perpendicular to this axis. In practice we are interested in particles with orbits not too far from this ideal.

The vertical excursions of the particles are governed by the radial component B_r of the magnetic field. One can see from Figure 1(b) that if the particles are moving in the sense that makes B_z hold them in an orbit centred on the z-axis, a field with $n<0$ gives forces tending to reduce any vertical displacement from the central plane, that is, tending to cause vertical focusing. On the other hand a field with $n>0$ will give vertical defocusing, any small vertical displacement leading ultimately to loss at the top or bottom of the vacuum chamber. It is therefore clear that an accelerator with a circularly symmetrical field must have $n<0$.

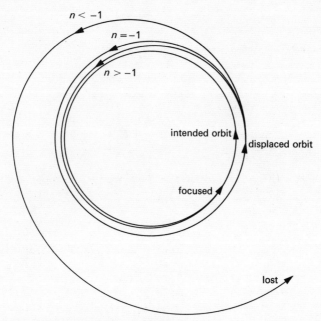

Figure 2 Horizontal focusing requires $n > -1$

2.2.3 Horizontal focusing

Radial excursions must be examined also, to discover the conditions for focusing in the horizontal plane. Equation **2.1**, considered as a relation between B_z and R, shows that if $n = -1$, possible orbits for particles of the same momentum p are concentric circles of different radii. Thus with $n = -1$, a particle which finds itself slightly outside its intended orbit will continue in a concentric orbit, with displacement neither increasing nor decreasing. This is the neutral condition between radial focusing and defocusing; defocusing occurs when $n < -1$, because then a particle displaced outwards sets off on part of a circular orbit of too large radius, and is eventually lost (see Figure 2). The condition for radial focusing, in which radial displacements are prevented from building up, is therefore

$$n > -1.$$

2.2.4 Summary

The two preceding sections, summarized in Table 1, show that there is a range of values of n,

$$-1 < n < 0,$$

22 Particle Accelerators

Table 1

Field index n	< -1	-1	0	>0
Vertical focusing?	Yes	Yes	No	
Radial focusing?	No	Yes	Yes	
Useful range		←——————→		

over which both vertical and radial focusing can be expected. Any accelerating machine using a circularly symmetric magnetic field must have a field index lying within these limits.

Betatrons, and the simpler cyclotrons and synchrocyclotrons mostly fall into this category, but it must be pointed out that there do exist special configurations of field designed to give stronger focusing. Some of these are discussed in sections 2.6 and 2.7.

2.3 The betatron

The betatron is an accelerator in a class by itself. It is discussed here because it is cyclic, depending on a magnetic field to hold the particles in circular orbits, and subject to the focusing conditions derived above for circularly symmetric fields. But there is no synchronization of accelerating voltage with the rotation of the particles, because there is no accelerating voltage, alternating or otherwise. Instead the acceleration is provided by the induced electric field around a changing magnetic flux. The basic theory of the betatron, as given in many elementary texts, shows that particles starting when the flux is zero are accelerated by an increasing flux through their orbit so that their momentum is always proportional to the flux

$$p = \frac{e}{2\pi rc} \times \text{flux}.$$

The magnetic field B needed to hold the particles in an orbit of constant radius r is proportional to the momentum, according to

$$B = \frac{pc}{re}$$

and therefore must increase in proportion to the flux through the orbit. Putting

$$B = \frac{c}{re}\frac{e}{2\pi rc} \times \text{flux} = \frac{\text{flux}}{2\pi r^2},$$

shows that the field B at the orbit must be half the mean value of the field inside.

23 The betatron

It is convenient that this condition is easily made compatible with the focusing condition that the field index should be between 0 and -1.

The final momentum achieved in a betatron of radius R, with maximum field at the orbit B_{max}, is

$$p_{max} = \frac{e}{c} r B_{max}.$$

Thus an accelerator of this type gives particles a final momentum independent of their mass. If the mass is large, this momentum may give a uselessly small kinetic energy, while the same momentum on a light particle gives a large kinetic energy. It is for this reason that acceleration by changing magnetic flux is useful for electrons only, and the machine based on it is called the betatron.

Many betatrons capable of accelerating electrons up to kinetic energies of order 20 MeV have been constructed as research tools. They are useful mainly as sources of high-energy γ-rays for the study of electromagnetic processes and photonuclear reactions.

2.4   The fixed-frequency cyclotron

In the simple type of cyclotron, an ion source produces protons, deuterons or other ions at the centre of a large electromagnet which gives a nearly uniform field over most of the flat cylindrical gap between the pole-pieces. The ion source may be a simple discharge tube, or a polarizing source of the type described in section 12.5. The normal slow decrease of field towards the outside of the gap is adjusted to give a field index appropriate for good focusing. A vacuum chamber encloses the gap between the pole-pieces, and within it are two D-shaped electrodes to which is fed the output of a radio-frequency generator. The frequency and the magnetic field are set to suit the value of e/m for the particles to be accelerated: the frequency f has to be given by

$$f = \frac{Be}{2\pi mc}. \qquad \qquad \textbf{2.2}$$

The alternating potential difference between the electrodes should pick up bunches of particles which can be accelerated through many cycles, with kinetic energy, momentum and radius of orbit increasing steadily.

The slow radial decrease of field needed for focusing has repercussions on the synchronization of the orbits with the accelerating voltage, for the frequency of rotation of the particles in their orbits decreases as they reach the weaker field in the outer part of the magnet. As a compromise, the radio frequency is usually made a little too low for perfect synchronization of the innermost orbits. This means that particles which start crossing the gap between the electrodes at peak accelerating voltage make the next crossings too early, to the left of the peak in Figure 3. After this phase excursion to the left, the particles reach the region where the field has dropped below the value for

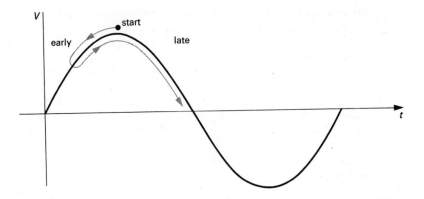

Figure 3 Phase of crossing gap for a particle being accelerated in a fixed-frequency cyclotron

perfect synchronization, and the representative point climbs back to the peak of Figure 3. As soon as the point starts to slide down the right-hand side of the peak, particles crossing early start to be accelerated more, and so to gain more momentum and move out to the region of still lower rotation frequency, which makes them arrive at the next crossing less early. This effect is known as phase-focusing or bunching. In the end, the particles all reach the phase at which they make their last crossings at zero potential difference. In practice the accelerated particles are usually extracted before this occurs, for example, by means of an electrostatic deflecting electrode placed at a predetermined radius.

Parameters of a few typical fixed-frequency cyclotrons are listed in Table 2, which includes also some sector-focusing instruments (see section 2.7).

2.5 The synchrocyclotron

In the simple cyclotron, inhomogeneity of the magnetic field leads to imperfect synchronization of the rotating particles with the accelerating potential difference. For magnets of moderate diameter, this effect usually leads to serious discrepancies of phase as the kinetic energy approaches 10 MeV (for protons). Adjustment of the field to restrain this effect is of little value in allowing acceleration to higher energies, because the relativistic increase of mass is already beginning to make a further similar discrepancy of phase. In the synchrocyclotron, or frequency-modulated cyclotron, the magnetic field index is chosen to give good focusing over a wide range of orbit radii, and problems of synchronization of phase, whether due to radial decrease of magnetic field or to relativistic increase of mass, are all met together by periodic decreasing of the frequency of the accelerating voltage.

25   The synchrocyclotron

Table 2 Parameters of Some Fixed-Frequency Cyclotrons

Laboratory	Magnet diameter /m	field /kG	Frequency /MHz	Focusing	Particle	Kinetic energy /MeV	Circulating current /μA
Argonne, USA	1·52	15	11·2	simple	α	43·2	600
					d	21·6	1000
					p, in H_2^+	10·8	600
Birmingham, UK	1·52	13·5	10·3	simple	α	40	25
					d	20	350
					p, in H_2^+	10	100
LRL Berkeley, USA	2·20	9	11	simple	α	36	1000
					d	18	
AERE Harwell, UK	1·78	17	7·6–22	3 sectors, spiral angle 45°	p	50	1000
Oak Ridge, USA	1·93	17	7·5–22·5	3 sectors, spiral angle 30°	p	10–70	500
Orsay, France	2·00	15	4–11	3 radial sectors	p	4–140	500

The peak value of the frequency is chosen to allow particles to start at the centre and be accelerated to a moderate energy as if in a simple cyclotron, with phase as shown in Figure 3. Then as the frequency begins to drop, the bunch of particles remains at a roughly constant phase near the axis on the right-hand side of the peak, each reduction in radio frequency allowing the phase to slip nearer to the peak, causing extra acceleration until the mass has increased enough to restore the synchronization.

In this way a single bunch of particles can be accelerated each time the frequency is reduced, up to an energy fixed by the radius of the magnet and the field strength at this radius, provided that the change of frequency is enough to allow for the whole change of field and mass as the particles go from orbits of zero radius to the orbit of maximum radius.

In practice the frequency is allowed to swing up and down in a roughly sinusoidal modulation. The half-cycle in which the frequency is rising is unused (see Figure 4), but this is not a serious disadvantage from the point of view of intensity. For example, a typical burst of 3×10^{10} protons every descending half-cycle, with a modulation frequency of 200 Hz, gives a mean circulating current of one microampere. This is a conveniently high intensity

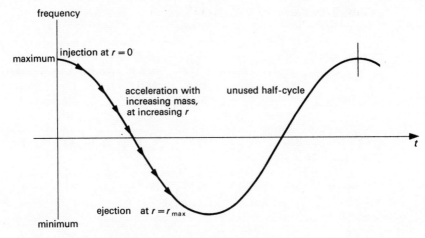

Figure 4 Frequency modulation in the synchrocyclotron

for many purposes, though of course there are experiments for which more is useful despite the embarrassment of half a kilowatt of beam energy ($1\mu A \times 500$ MeV) and the general radioactivity which it can produce.

The frequency modulation is obtained by including a continuously variable capacity in the L–C resonant circuit controlling the radio-frequency oscillator. The two commonest devices for varying capacity with the required modulation frequency are vibrating tuning forks and rotating condenser plates.

Parameters for a few existing synchrocyclotrons are listed in Table 3.

Table 3 Parameters of Some Synchrocyclotrons

Laboratory	Proton kinetic energy /MeV	Magnet		Radio frequency			Circulating current /μA
		diameter /m	maxi- mum field/kG	range /MHz	voltage /kV	modula- tion fre- quency /Hz	
LRL Berkeley, USA	740	4·77	23·3	36–18	10	64	1
Dubna, USSR	680	6·00	16·8	26–13	15	120	2·4

Table 3 – *continued*

Laboratory	Proton kinetic energy /MeV	Magnet		Radio frequency			Circulating current /μA
		diameter /m	maximum field/kG	range MHz	voltage /kV	modulation frequency /Hz	
CERN, Switzerland	600	5·00	19·5	30–16·4	30	54	1·8
Chicago, USA	460	4·30	18·6	28–18	14	85	1
Orsay, France	160	2·80	16·3	25–20	25	450	4
AERE, Harwell, UK	160	2·70	16·2	26–19	10	200	1·5

(Note: the 410 MeV Liverpool synchrocyclotron ceased operation in 1967.)

2.6 The synchrotron

2.6.1 *Principles*

To accelerate protons to kinetic energies approaching 1 GeV (1000 MeV or 10^9 eV) in a synchrocyclotron would require a very large and heavy magnet, and it turns out that the synchrotron is more practical for this task. In the synchrotron, particles are accelerated in an orbit of constant radius, instead of in a spiral. This allows the use of a ring-shaped magnet, generating field in an annular region, the centre being completely empty.

The conditions for obtaining constant radius are that the magnetic field should increase in proportion to the momentum of the protons (equation **2.1**), and that the accelerating frequency should be kept equal to the frequency of rotation, which increases in proportion to B/m (equation **2.2**). With the relativistic value of the mass included, this gives for the relation between frequency and magnetic field

$$f = \frac{e}{2\pi m_0 c} B \left[1 + \left(\frac{BeR}{m_0 c^2} \right)^2 \right]^{-\frac{1}{2}}, \qquad \textbf{2.3}$$

where m_0 is the rest mass of the proton and R the radius of the orbit.

In practice the magnetic field is allowed to rise at a rate set by the self-inductance of the magnet and the characteristics of the generating machinery

28 Particle Accelerators

from which it is supplied. The accelerating frequency is then made to follow the field, according to equation **2.3**, which is embodied in a preset programme. Electrodes to detect the radial position of the beam may be used for fine control of the frequency. The whole cycle takes a few seconds in most machines, one exception being the Princeton–Pennsylvania accelerator in which the magnet forms part of a resonant circuit giving nineteen cycles per second.

Neither the magnetic field nor the frequency can be controlled accurately at the beginning of the cycle, when both should be near zero, so protons are injected into the main orbit after the cycle has started, from a smaller auxiliary accelerator, which may be a van de Graaff or a linear accelerator. The injection energy chosen is normally higher than is necessary for overcoming the initial irregularities of the magnetic field, because high injection energy reduces the space–charge limitation on beam intensity. Injectors for the largest proton synchrotrons are therefore linear accelerators producing energies of order 50 MeV.

The radio-frequency accelerating voltage, instead of being fed to D-shaped or C-shaped electrodes, is used to drive one or more cylindrical cavities enclosing part of the orbit, at a frequency which is a harmonic of the frequency of rotation of the particles.

The first proton synchrotrons were built with constant-gradient focusing or weak focusing, according to the ideas outlined in section 2.2: the magnet pole-pieces were shaped to give a field index between 0 and -1 over a vacuum chamber large enough to accommodate the expected vertical and radial oscillations. For kinetic energies up to 12·5 GeV this proved satisfactory, though magnet weights became rather large (7000 tons for Nimrod, producing 7 GeV).

2.6.2   *Alternating gradient focusing*

To go to energies above 12·5 GeV, strong focusing based on an alternating gradient of magnetic field has been adopted universally since the idea was proposed by Courant, Livingston and Snyder (1952). This has allowed the wide beams of weak-focusing machines to be replaced by pencil beams which have great advantages for experimental work, in addition to the economies which they allow in the size of the vacuum chamber and hence in weight and cost of the magnet.

The whole technique of focusing beams of charged particles is limited by the fact that a transverse magnetic field which gives convergence in one plane causes divergence in the other, as we have already shown for the radially decreasing field of a cyclotron. Thus a single source of inhomogeneous field, whether it is the edge of a bending magnet giving non-zero mean field, or a quadrupole magnet with zero mean field designed for focusing without bending, has an opposite effect in two perpendicular planes, usually vertical and horizontal or $\pm 45°$ according to the construction. The analogy between beam optics and ordinary optics is thus incomplete: within a single transverse plane

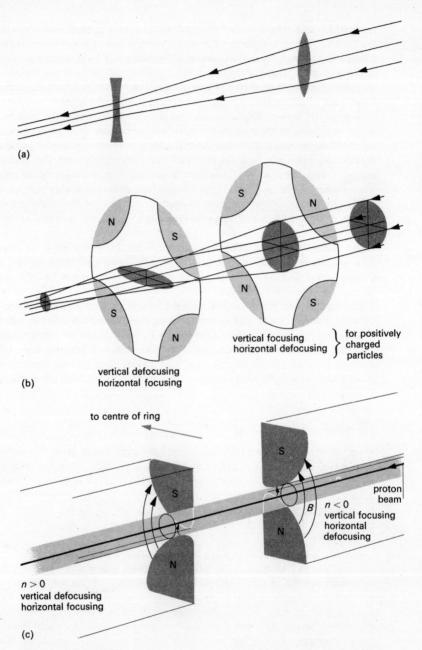

(a)

S

N

N

S

S

N

vertical focusing
horizontal defocusing

} for positively
charged
particles

vertical defocusing
horizontal focusing

(b)

to centre of ring

S

S

N

N

B

proton
beam

$n < 0$
vertical focusing
horizontal
defocusing

$n > 0$
vertical defocusing
horizontal focusing

(c)

Figure 5 Alternating-gradient focusing. (a) Optical; (b) quadrupole lenses; (c) in AG synchrotron

30   Particle Accelerators

it is satisfactory, but what happens to particles in two perpendicular transverse planes has to be represented by an optical system in which all lenses have cylindrical curvature as well as spherical. Fortunately for the optical industry, glass lenses have no such limitations.

However, in an accelerator we are not usually concerned with forming a faithful image of a source. Rather, we need to limit the tendency of a nearly parallel beam of particles to spread out and fill too large a volume. In the case of an optical system, this can be done with a series of alternate converging and diverging lenses. The corresponding effect is obtained for charged particles with a series of magnets giving inhomogeneous fields with gradients of alternate sign (see Figure 5). While this device is unnecessary in optical instruments, it is useful in particle beams for giving a focusing effect in both transverse planes. For example an individual quadrupole magnet may give vertical focusing and horizontal defocusing; but made into a pair with an opposite quadrupole of similar strength placed behind it, or made into a triplet with two others of half its strength, it can give focusing in both planes.

In a synchrotron this effect is obtained by constructing the ring magnet out of a large number of separate magnets with alternately positive and negative values of the field index n. For the formal theory of the orbits obtained with an alternating gradient of this type, the student is referred to the treatment given by Farago (1970). The practical result has been the construction of proton synchrotrons with vacuum chambers and magnets much smaller than would have been necessary for weak-focusing machines of similar energy. For example, the 28 GeV alternating-gradient CERN proton synchrotron has a total magnet weight of 3500 tons, half that of the magnet used in the weak-focusing Nimrod for protons of only 7 GeV. While these figures show the saving resulting from use of the alternating-gradient principle, the advantage of the synchrotron over the synchrocyclotron is shown by comparing them with the figure of 1600 tons, the weight of the magnet of the CERN synchrocyclotron, which accelerates protons to the much smaller energy of 600 MeV.

2.6.3 *Actual proton synchrotrons*

The parameters of some existing and proposed proton synchrotrons are listed in Table 4, which is divided into sections showing (a) the weak-focusing machines covering energies up to 12·5 GeV, and (b) the alternating-gradient machines which are operating at energies up to 70 GeV and are projected for 200 GeV in the USA and for 300 GeV in Europe.

It may also be mentioned that two of the earliest proton synchrotrons, namely the 3 GeV Brookhaven Cosmotron and the 1 GeV Birmingham synchrotron, have already ceased operation.

2.6.4 *Extraction*

Synchrotrons are often used by raising a target into the circulating beam at the end of each acceleration cycle. Beams of secondary particles from interactions

Table 4 Parameters of Some Proton Synchrotrons

Laboratory	Name	Energy/GeV kinetic	total	Mean radius /m	Injection	Frequency range/MHz	Harmonic used	Cycle time/s	Years of operation
Birmingham, UK	—	1	2	4·5	0·46 MeV v.d. Graaff	0·33–9·3	1	10	1953–67
Brookhaven, USA	Cosmotron	3	4	9	3·6 MeV v.d. Graaff	0·36–4	1	5	1952–66
Saclay, France	Saturne	2·5	3·5	11	3·6 MeV v.d. Graaff	0·78–8·41	2	3	1958–
Princeton, USA	PPA	3	4	12·2	3 MeV v.d. Graaff	2·5–30	8	0·052	1960–
Berkeley, USA	Bevatron	6	7	18·2	20 MeV linac	0·36–2·46	1	5	1954–
RHEL, Chilton, UK	Nimrod	7	8	23·6	15 MeV linac	1·4–8·0	4	2–3	1961–
JINR, Dubna, USSR	Synchro-phasotron	10	11	30·5	9 MeV linac	0·19–1·45	1	20	1957–

Constant gradient

Argonne, USA	ZGS	12·5	13·5	27·5	50 MeV linac	4·4–14	8	4	1962–
CERN, Switzerland	CPS	27	28	100	50 MeV linac	2·9–9·55	20	3·5	1959–
Brookhaven, USA	AGS	32	33	128·5	50 MeV linac	1·4–4·5	12	3	1960–
Serpukhov, USSR			70	235	100 MeV linac		30	6–12	1967–
NAL, Batavia, USA		200/500	1000		200 MeV linac – 8 GeV booster				Proposed 1972–
Europe		300+	1100		10 GeV from CPS				Proposed 1975–

Alternating gradient

in the target may then be transported to experimental areas along beam lines in which quadrupole magnets, bending magnets and sometimes electrostatic separators serve to select particles of the nature and momentum required for the particular experiment. Several such secondary beams may operate simultaneously from a single target, and it is possible to share the circulating beam between several targets. Thus a single accelerator may serve many experiments at the same time.

However, more flexibility is obtained if the proton beam itself can be extracted and carried to an experimental area. Many schemes exist for extracting synchrotron beams: most depend on using some type of pulsed magnet to displace part of the circulating beam out of its stable orbit. Then, at the peak of the oscillation resulting from this displacement, a second magnet catches the displaced particles and deflects them still further, out of the influence of the focusing magnetic field and into an external beam line.

2.7 The sector-focusing cyclotron

In the preceding sections we have discussed the need for alternating-gradient focusing in synchrotrons for accelerating protons to the highest energies. However, this is not the only field in which something better than constant-gradient focusing is needed; ordinary fixed-frequency cyclotrons can be made to yield higher beam currents, and to operate up to higher energies, by introducing azimuthal variations of magnetic field which can give focusing in both planes even when the mean field index n is positive.

These azimuthal variations may be obtained by using pole-pieces with radial ridges; the theory then is very similar to that of the alternating-gradient synchrotron. Still further focusing is obtained by making the ridges spiral: then the particles move obliquely across the boundaries between low-field sectors and high-field sectors, and experience focusing forces determined by the angle of the spiral.

The full theory of sector-focusing will be found in the literature (e.g. review by Richardson, 1965). The parameters of three sector-focusing cyclotrons are included in Table 2.

Chapter 3
Kinematics

3.1 Types of nuclear process

In this chapter we shall be examining the limitations placed upon the motion of colliding particles by the laws of mechanics, in particular the laws of conservation of energy and momentum. We shall assume a given state of motion for two particles about to collide and shall study the motion of the products of the collision.

As a generalized nuclear process, we consider a collision in which particle 1, moving with momentum $\mathbf{p}_1$, strikes a previously stationary particle 2, to produce an unspecified number of final particles f: we may give each of the final particles a momentum $\mathbf{p}_f$, a kinetic energy T_f, and a rest mass M_f; the initial particles are similarly given rest masses M_1 and M_2, and kinetic energies T_1 and 0 respectively.

If the final particles are the same as the two initial particles, the process is called *scattering*. The scattering is said to be *elastic* if no energy is used in raising either of the particles to an excited state.

If the product particles are different from the initial, the process is called a *reaction*. When there are only two of them, it is called a two-body reaction. In some cases there may be only one product particle of non-zero rest mass; this type of process is known as a *capture* reaction.

The principles to be laid down here will apply to all these types of process, though most of the calculations will limit the number of final particles to two. The conclusions will therefore be applicable to elastic scattering, inelastic scattering (considered as a reaction in which one final particle is an excited state of one of the initial particles), and capture reactions (the second final particle being a γ-ray photon in the case of radiative capture). By slight extension of the conclusions, they may also be used to cover the spontaneous decay of an unstable particle: if we omit particle 1, we have the rules for decay at rest; alternatively if we omit particle 2, the process described is the decay in flight of particle 1.

3.2 Conservation of energy

3.2.1 *Stored energy*

In nuclear processes it is particularly easy to trace the operation of the law of conservation of energy: as in all other types of process, the quantity conserved is the total energy of the system, that is, the sum

Stored energy, internal to the component particles + kinetic energy of motion of the component particles + contributions to the energy of the system from the outside world.

In a chemical process, the stored energy may be calculated only through a series of book-keeping operations: at first it may look like a fraudulent device, introduced solely to preserve the law of conservation of energy. The fact that it is not fraudulent is demonstrated only when release of the stored energy shows that the book keeping leads to correct predictions.

If we could measure masses accurately enough, we could dispense with the book keeping, because the energy stored in an object is reflected in its mass, according to the Einstein relation

$$E = mc^2.$$

We may, if we like, express the mass of a particle in units of energy, and regard it all as stored energy. This device tells us how much the mass of a particle will change when a given amount of energy is absorbed or emitted. But it tells us nothing about how large a proportion of the total mass can be released as energy; this proportion depends on circumstances.

3.2.2   *Energy balance in reactions*

In mechanical processes objects can be given energies equivalent to only about 10^{-10} of their total mass; in chemical processes the factor is still not more than about 10^{-9}. Because these proportions are so small, it is usually easier in these fields to think of a constant mass and a variable stored energy as separate properties.

In a nuclear reaction the energy change may be equivalent to a mass of order 10^{-3} of the total mass. There remains a basic mass which is not convertible into energy by any known process,† but it is only a thousand times greater than the additional mass representing convertible stored energy. The unconvertible basic mass represents the unchanging total number of nucleons (i.e. neutrons + protons), while the additional (positive or negative) mass represents (positive or negative) stored energy appropriate to the way the nucleons are bound in the nucleus under consideration. We choose as unit of mass, not the mass of a free nucleon which has proportionately more stored energy than any other nucleon, but a nucleon in a standard nucleus, namely $^{12}_{6}C$. To be precise, we take one-twelfth of the mass of a $^{12}_{6}C$ atom, so that in taking the mean mass of neutron and proton we include an electron with the proton. Then we get atomic masses, for individual isotopes, which are all within 0·1 of the integer representing the number of nucleons present. The small excess represents the additional stored energy, which is negative for all but the smallest and the largest nuclei, following the trend illustrated in Figure 34(b) (p. 135). If the excess is expressed in mass units, it may be converted to energy by the relation

†Except the annihilation of a nucleon–antinucleon pair, which in fact only releases energy equivalent to that previously used up in creating the pair.

1 mass unit (on scale $^{12}_{6}C = 12$) = 931·44 MeV.

Thus if the mass is specified to the nearest thousandth of a mass unit, the stored energy is known to the nearest million electronvolts. Many nuclear masses are known within 10^{-6} mass unit, which fixes the energy within a thousand electronvolts.

Some writers prefer to write MeV/c^2 as the unit of mass equivalent to the MeV of energy. This is a device for reminding the reader that the quantity being expressed in millions of electronvolts is mc^2 rather than m itself, and that certain calculations† will require insertion of the factor c^2.

3.2.3 *Q-value of a reaction*

In a reaction an amount of energy Q may be released, that is, converted from energy stored in the masses of the particles to kinetic energy of motion. Conservation of energy then requires

$$\sum T_f = T_1 + Q \qquad \qquad \textbf{3.1}$$

and

$$Q = M_1 + M_2 - \sum M_f.$$

Q, which may be positive, zero, or negative, has come to be known as the Q-value of the reaction.

Accurate measurements of Q-values of reactions have played a valuable part in establishing self-consistent tables of nuclear masses, such as that shown in the Appendix.

3.3 Conservation of momentum

The general requirement for conservation of momentum, in the notation of section 3.1, is

$$\mathbf{p}_1 = \sum \mathbf{p}_f, \qquad \qquad \textbf{3.2}$$

where $\sum \mathbf{p}_f$ means the vector sum of all final momenta.

For some purposes, however, it may be convenient to resolve the final momenta into longitudinal components p_x, in the direction of $\mathbf{p}_1$, and transverse components p_y and p_z in two directions at right angles to $\mathbf{p}_1$. We then have three conditions:

$$\left. \begin{array}{l} p_1 = \sum p_x, \\ 0 = \sum p_y, \\ 0 = \sum p_z. \end{array} \right\} \qquad \qquad \textbf{3.3}$$

†As an exercise, work out the mass equivalent to 1 MeV, and show that the result is the same whether you use c.g.s. or SI units.

Each momentum may be expressed as the product mv of a mass with a velocity. If the velocity is non-relativistic, the mass m is just the rest mass M, and the momentum is related to the kinetic energy by

$$p = mv = Mv = \sqrt{(2MT)}. \qquad \textbf{3.4}$$

But if the velocity is relativistic, we must use the more general relations

$$m = \frac{M}{\sqrt{(1 - v^2/c^2)}},$$

$$E^2 = p^2c^2 + M^2c^4,$$

where E = total energy = $Mc^2 + T$.

\qquad \textbf{3.5}

The kinematic rules for the process are obtained by simultaneous solution of the equations for momentum balance **3.2** or **3.3** and the energy equation **3.1**, with kinetic energies and momenta related by equation **3.4** or **3.5**. Complete solution is not usually possible, since there are more unknowns than equations; but the equations provide relations between the unknowns. In the following section we consider the case of two final particles, for which these relations allow us to specify everything about the final system, in terms of one of its parameters.

3.4 Two-body processes

3.4.1 *The general method*

Let us consider a simple two-body process in which particle 1 collides with a stationary particle 2 to produce two final particles 3 and 4. This limitation to two final particles greatly simplifies the kinematical calculations, which we shall do first under non-relativistic (Newtonian) conditions and then by methods which are also valid under relativistic conditions.

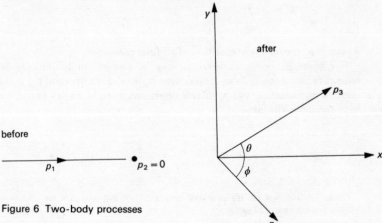

Figure 6 Two-body processes

The conclusions should apply to a nuclear reaction with two products, and, by putting $Q = 0$ with particles 3 and 4 the same as particles 1 and 2, we can make them describe elastic scattering.

The first implication resulting from limiting the final particles to two is that the directions of $\mathbf{p}_3$ and $\mathbf{p}_4$ define a plane of production. We may take our y-axis in this plane, so that all z-components of momentum are zero (see Figure 6). The three equations **3.3** for conservation of the three components of momentum are now reduced to two, namely

$$p_1 = p_3 \cos \theta + p_4 \cos \phi, \qquad\qquad\qquad \textbf{3.6}$$

$$0 = p_3 \sin \theta - p_4 \sin \phi. \qquad\qquad\qquad \textbf{3.7}$$

Now we must introduce the condition for conservation of energy, and this is where the choice of argument depends on whether or not the velocities are liable to be relativistic.

3.4.2 *Non-relativistic treatment*

In non-relativistic conditions we put

$$T_1 + Q = T_3 + T_4$$

and express the kinetic energies in terms of the momenta to give

$$\frac{p_1^2}{2M_1} + Q = \frac{p_3^2}{2M_3} + \frac{p_4^2}{2M_4}. \qquad\qquad\qquad \textbf{3.8}$$

Now we have three equations, **3.6–8**, relating four unknowns, p_3, p_4, θ and ϕ. We can assume Q and p_1 are known. We can therefore eliminate any two unknowns, leaving a relation between the other two. Let us start by eliminating p_4 and ϕ to obtain p_3 in terms of θ.

From equations **3.6** and **3.7** we eliminate ϕ by squaring and adding:

$$p_4^2 = p_4^2(\cos^2\phi + \sin^2\phi) = (p_1 - p_3 \cos \theta)^2 + p_3^2 \sin^2\theta$$

$$= p_1^2 - 2p_1 p_3 \cos \theta + p_3^2.$$

If we substitute this value for p_4^2 in equation **3.8**, we get

$$\frac{p_3^2}{2M_3} + \frac{1}{2M_4}(p_1^2 - 2p_1 p_3 \cos \theta + p_3^2) = \frac{p_1^2}{2M_1} + Q.$$

Hence $\quad p_3^2\left[\dfrac{1}{M_3} + \dfrac{1}{M_4}\right] - p_3 \dfrac{2p_1 \cos \theta}{M_4} + p_1^2\left[\dfrac{1}{M_4} - \dfrac{1}{M_1}\right] - 2Q = 0. \qquad \textbf{3.9}$

This is a quadratic in p_3, giving (according to conditions) two, one or no positive values of p_3 for each value of θ. Thus we may expect to find, at a given angle θ, up to two groups of final particles of type 3, with well defined momenta.

If the condition

$$\frac{Q}{T_1} \geqslant \frac{M_1}{M_4} - 1$$ 3.10

is satisfied, there is one group of particles at every value of the angle θ. But for a single angle θ, the condition for finding particles is less rigorous:

$$\frac{Q}{T_1} \geqslant \frac{M_1}{M_4} \cdot \frac{(M_3 \sin^2\theta + M_4)}{(M_3 + M_4)} - 1.$$ 3.11

Over the angles for which condition **3.11** holds but condition **3.10** does not there will be two groups of particles. Where neither condition is satisfied, no particles of type 3 will be observed.

Having obtained equation **3.9** as the relation between the direction and the momentum for one of the two final particles, we may substitute in equations **3.8** and **3.7** to find the direction and momentum of the other final particle. The physically interesting point about a two-body process is that when we have specified the momentum or the direction of one of the final particles, everything else is fixed.

3.4.3 Relativistic treatment

Here we use equations **3.6** and **3.7** in conjunction with a relativistic equation for conservation of total energy,

$$E_1 + M_2 c^2 = E_3 + E_4.$$ 3.12

E_1, E_3 and E_4 are the total energies of the three particles, and are related to the corresponding momenta by equation **3.5**. Since particle 2 is at rest, its total energy is its rest energy $M_2 c^2$.

The general simultaneous solution of equations **3.12**, **3.6** and **3.7**, with energies and momenta related by equation **3.5** is clumsy. Even in the special case of elastic scattering the algebra provides more labour than insight. Since relativistic treatment is not necessary for the description of most ordinary nuclear processes, a full treatment is left to a companion volume (Hughes, 1971), where it is needed for describing the interactions of elementary particles at high energies.

Nevertheless, the principles are outlined, with simple examples, in section 3.5.4, as an illustration of how to use the centre-of-mass system.

3.5 Centre-of-mass system

3.5.1 Centre of momentum

So far we have considered the kinematics of collisions and reactions, as they appear to a stationary observer in an ordinary laboratory. The set of axes and the clock used by this sort of observer are said to provide the laboratory system of coordinates.

But it is sometimes convenient to study a process as it would appear with respect to a set of axes moving with the same velocity as the centre of mass of the interacting particles. This set of axes provides the 'centre-of-mass' system of coordinates, which would be better called 'centre-of-momentum' for the following reason: the centre of mass of a set of particles is the point about which their masses have zero moment. What we want is a system in which the total momentum is zero. It is sufficient to describe this system by a velocity vector, without specifying the actual position of its origin. Since this velocity is also the velocity of the centre of mass, the interchangeability of the names is explained. In practice, the abbreviation c.m.s. is used.

3.5.2 Two-body collision in c.m.s.

In the centre-of-mass system, a collision looks like two particles moving towards each other, with equal and opposite momenta. They interact, and if it is a two-body process the two final particles are emitted with equal and opposite momenta, along some line making an angle θ_c with the line of motion before the collision. If the process is elastic scattering, the final momenta are equal in magnitude to the initial momenta. But in other processes the final momenta will in general differ from the initial momenta. They will have values such that the final particles have a total kinetic energy equal to the total initial kinetic energy plus any energy Q released in the reaction.

3.5.3 Energy in c.m.s.

Let us take a particle of mass m_1, incident with non-relativistic velocity v and kinetic energy $T_1 = \frac{1}{2}m_1 v^2$ on a stationary particle of mass m_2. In the centre-of-mass system the two particles will be approaching each other with respective velocities v'_1 and v'_2. With the appropriate masses, these velocities must give the particles equal momenta in the centre-of-mass system,

$$p' = m_1 v'_1 = m_2 v'_2. \tag{3.13}$$

Being oppositely directed, they must also add to give a relative velocity v:

$$v'_1 + v'_2 = v. \tag{3.14}$$

Elementary algebra shows that the total incident kinetic energy in the centre-of-mass system is then

$$E = \tfrac{1}{2}m_1 v'^2_1 + \tfrac{1}{2}m_2 v'^2_2$$

$$= \tfrac{1}{2}p'v$$

$$= \tfrac{1}{2}mv^2 \tag{3.15}$$

$$= \frac{p'^2}{2m}, \tag{3.16}$$

where m is the reduced mass given by

$$m = \frac{m_1 m_2}{m_1 + m_2}.$$ 3.17

Thus the relative velocity v, the c.m.s. momentum of each particle, the reduced or relative mass m and E, which is sometimes called the relative energy, are all related in the same way as the velocity, the momentum, the mass and the kinetic energy of a single particle. The simplification of algebra which follows from this is used in many two-body calculations (see, for example, sections 10.2.2 and 11.4).

To link these parameters with those observed in the laboratory, a convenient relation follows from equations 3.15 and 3.17, which may be combined to give

$$E = \frac{m_2}{m_1 + m_2} T_1.$$ 3.18

3.5.4 The relation between velocities in the two systems
When two particles of equal momentum have different masses, their velocities are inversely proportional to their masses (equation 3.13). The velocity vectors in a two-body reaction, as seen in the centre-of-mass system, therefore look like Figure 7(a). Seen from the laboratory, however, the velocities of the final

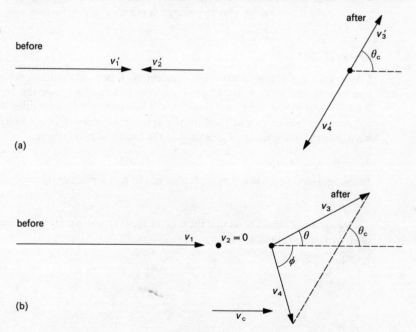

Figure 7 Velocities in two-body processes: (a) in c.m.s.; (b) in laboratory

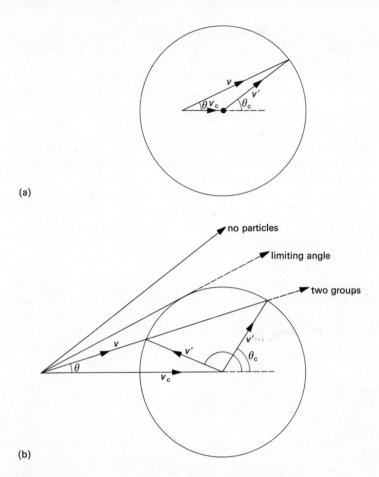

(a)

(b)

Figure 8 Relation between c.m.s. angle θ_c and laboratory angle θ. The velocity v_c is that of the c.m.s.

particles are as shown in Figure 7(b), which is the same as Figure 7(a) with the velocity of the centre-of-mass added.

Figures 8(a) and 8(b) illustrate the three cases discussed in section 3.4.2. Figure 8(a) is for the case in which only one group of particles is produced at a given angle θ. Figure 8(b) is for the case in which, for angles up to a certain limit, there are at each angle θ two groups with different values of velocities. Beyond the angle at which these groups coalesce there are no particles at all; this is the third case of section 3.4.2.

In non-relativistic conditions, the velocity of the centre-of-mass is added by the ordinary laws of vector addition. There the condition for using Figure 8(a) is the inequality **3.10**. When this does not hold, Figure 8(b) is appropriate, with its limiting angle given by the inequality **3.11**. But in relativistic conditions one must use relativistic rules for addition of velocities; alternatively, one may reach equivalent conclusions without reference to the velocities, by applying relativistic transformations to the momenta and energies.

3.5.5   *C.M.S. treatment of relativistic two-body reaction*

Use of the centre-of-mass system simplifies the relativistic treatment of two-body reactions.

Let us suppose that the incident particle has ordinary three-momentum $\mathbf{p}_1$, directed along the x-axis, so that with its total energy E_1 it has a four-momentum with components $(p_1, 0, 0, iE_1/c)$. The target particle has zero three-momentum, but its total energy E_2 is $M_2 c^2$. The system as a whole therefore has:

total three-momentum $\mathbf{P} = \mathbf{p}_1$,

total energy $E = E_1 + M_2 c^2$.

It may therefore be considered as a composite object with rest mass M^* and velocity $\beta_c c$, where $M^* c^2$ is the total energy measured in the c.m.s., and β_c is the velocity of the centre of mass, expressed in units of c. These quantities are given in terms of the total three-momentum $\mathbf{P}$ and total energy E by

$$\beta_c = \frac{P_c}{E} = \frac{p_1 c}{E_1 + M_2 c^2} \qquad \textbf{3.19}$$

and $(M^* c^2)^2 = E^2 - P^2 c^2$

$$= (E_1 + M_2 c^2)^2 - p_1^2 c^2.$$

Thus
$$M^{*2} = 2E_1 M_2 c^{-2} + (M_1^2 + M_2^2). \qquad \textbf{3.20}$$

The gamma factor for the centre of mass is

$$\gamma_c = (1 - \beta_c^2)^{-\frac{1}{2}} = \frac{E_1 + M_2 c^2}{M^* c^2}. \qquad \textbf{3.21}$$

The next step is to calculate the momenta of the final particles in the centre-of-mass system. To do this we liken the process to the decay at rest of a particle of rest mass M^* into two particles with rest masses M_3 and M_4. These particles must move off in opposite directions, with equal momenta p' such that the sum of their total energies is $M^* c^2$; this condition is written

$$\sqrt{(M_3^2 c^2 + p'^2)} + \sqrt{(M_4^2 c^2 + p'^2)} = M^* c.$$

The solution, obtained by squaring twice, is

$$p'^2 = \tfrac{1}{4} c^2 \{ M^{*2} - 2(M_3^2 + M_4^2) + (M_3^2 - M_4^2)^2 M^{*-2} \}. \qquad \textbf{3.22}$$

If the particles are moving in directions defined by a c.m.s. angle θ_c (see Figure 7a), their c.m.s. momenta will have components

$$p'_{3x} = p' \cos \theta_c, \qquad p'_{3y} = p' \sin \theta_c;$$

$$p'_{4x} = -p' \cos \theta_c, \qquad p'_{4y} = -p' \sin \theta_c.$$

In the c.m.s., particle 3 has total energy given by

$$E'_3 = \sqrt{(p'^2 c^2 + M_3^2 c^4)}$$

$$= \tfrac{1}{2} c^2 \left[M^* + \frac{M_3^2 - M_4^2}{M^*} \right]. \qquad \qquad \textbf{3.23}$$

Similarly, particle 4 has total energy in the c.m.s.

$$E'_4 = \tfrac{1}{2} c^2 \left[M^* + \frac{M_4^2 - M_3^2}{M^*} \right]. \qquad \qquad \textbf{3.24}$$

To obtain the momenta p_3 and p_4 in the laboratory system, we must use the appropriate Lorentz transformation. Expressed in terms of four-vectors and a transformation matrix, this is, for particle 3,

$$\begin{bmatrix} p_{3x} \\ p_{3y} \\ p_{3z} \\ \dfrac{iE_3}{c} \end{bmatrix} = \begin{bmatrix} \gamma & 0 & 0 & -i\gamma_c \beta_c \\ 0 & 1 & 0 & 0 \\ 0 & 0 & 1 & 0 \\ i\gamma_c \beta_c & 0 & 0 & \gamma \end{bmatrix} \begin{bmatrix} p'_{3x} \\ p'_{3y} \\ p'_{3z} \\ \dfrac{iE'_3}{c} \end{bmatrix}$$

The three components of momentum are therefore

$$\left. \begin{aligned} p_{3x} &= \gamma_c \left[p'_{3x} + \beta_c \frac{E'_3}{c} \right] = \gamma_c \left[p' \cos \theta_c + \beta_c \frac{E'_3}{c} \right], \\ p_{3y} &= p'_{3y} = p' \sin \theta_c, \\ p_{3z} &= 0. \end{aligned} \right\} \qquad \textbf{3.25}$$

These may be obtained, for a particular value of θ_c, by inserting the values of p' and E'_3 from equations **3.22** and **3.23**.

The direction θ of particle 3 in the laboratory system is given by

$$\cot \theta = \frac{p_{3x}}{p_{3y}} = \gamma_c \left(\cot \theta_c + \frac{\beta_c E'_3}{p'c} \operatorname{cosec} \theta_c \right), \qquad \textbf{3.26}$$

The particle 4, moving in a direction at an angle $\pi - \theta_c$ on the other side in the c.m.s., will have a laboratory angle ϕ given by

$$\cot \phi = \gamma_c \left(-\cot \theta_c + \frac{\beta_c E'_4}{p'c} \operatorname{cosec} \theta_c \right). \qquad \textbf{3.27}$$

3.6 Elastic scattering

Elastic scattering is an especially simple process which may be treated from first principles, or by putting into the equations for a general two-body process the conditions

$$M_3 = M_1; \qquad M_4 = M_2; \qquad Q = 0.$$

These conditions describe a collision in which particle 1 is incident and after being scattered is called particle 3, while particle 2 is the stationary target particle which when recoiling from the impact is called particle 4.

Either method may be used to give results which, expressed in terms of ϕ, the angle of the recoil particle, take the following form for non-relativistic conditions:

$$\text{Momentum of recoil particle} = p_4 = \frac{2M_2\, p_1 \cos \phi}{M_1 + M_2}. \qquad \textbf{3.28}$$

$$\text{Momentum of scattered particle} = p_3 = p_1 \sqrt{\left[1 - \frac{4M_1 M_2}{(M_1 + M_2)^2} \cos^2 \phi \right]}. \qquad \textbf{3.29}$$

The direction θ of the scattered particle is given by

$$\cot \theta = \frac{M_1}{M_2} \operatorname{cosec} 2\phi - \cot 2\phi. \qquad \textbf{3.30}$$

The kinetic energy of the recoil particle is given by

$$T_4 = \frac{4M_1 M_2}{(M_1 + M_2)^2} T_1 \cos^2 \phi, \qquad \textbf{3.31}$$

while that of the scattered particle is

$$T_3 = T_1 - T_4$$
$$= \frac{(M_1 - M_2)^2 - 2M_1 M_2 \cos 2\phi}{(M_1 + M_2)^2} T_1. \qquad \textbf{3.32}$$

It is often convenient to express these relations in terms of the c.m.s. angle of scattering θ_c. The conversion between the c.m.s. and laboratory systems is particularly simple for elastic scattering, because the c.m.s. velocity of the recoil particle is equal in magnitude to the velocity of the c.m.s. with respect to the laboratory. It follows (see Figure 8) that the c.m.s. angle of scattering θ_c and the laboratory-system angle of recoil are related by

$$\theta_c = \pi - 2\phi. \qquad \textbf{3.33}$$

From equation **3.30**, the laboratory-system angle of scattering is given by

$$\cot \theta = \frac{M_1}{M_2} \operatorname{cosec} \theta_c + \cot \theta_c. \qquad \textbf{3.34}$$

The laboratory-system energy of the recoil particle is given by equations **3.31** and **3.33** as

$$T_4 = \frac{2M_1 M_2}{(M_1 + M_2)^2} T_1(1 - \cos \theta_c). \qquad\qquad \textbf{3.35}$$

Correspondingly, equations **3.32** and **3.33** give for the laboratory-system energy of the scattered particle

$$T_3 = \frac{M_1^2 + M_2^2 + 2M_1 M_2 \cos \theta_c}{(M_1 + M_2)^2} T_1. \qquad\qquad \textbf{3.36}$$

This expression shows that the kinetic energy of the scattered particle, expressed as a fraction of the incident kinetic energy, ranges from 1 when $\theta_c = 0$ to a fraction.

$$\left(\frac{T_3}{T_1}\right)_{min} = \left(\frac{M_1 - M_2}{M_1 + M_2}\right)^2 \quad \text{when} \quad \theta_c = \pi. \qquad\qquad \textbf{3.37}$$

This result is of value in discussing the moderation of neutrons (see section 8.3).

3.7 Elastic scattering with equal masses

If particles 1 and 2 have equal masses, and undergo elastic scattering so that particles 3 and 4 are identical with them, the kinematic formulae become especially simple.

Important cases satisfying this condition are (a) the scattering of alpha particles in helium gas, (b) the scattering of protons by protons in hydrogenous material, and (c) the scattering of neutrons by protons, in so far as the slight difference in the masses can be neglected.

All of these are covered by the calculations which follow in sections 3.7.1 and 3.7.2. We put all the masses equal to a common mass

$$M = M_1 = M_2 = M_3 = M_4$$

and we note that there is no kinematic distinction between the recoil and the scattered particle, although the words are retained for convenience.

3.7.1 *Non-relativistic calculation*

With equal masses, equation **3.28** for the momentum of the recoil particle becomes

$$p_4 = p_1 \cos \phi.$$

Equation **3.34**, relating the directions of the scattered and recoil particles, reduces to

$$\cot \theta = \tan \phi,$$

from which we deduce that the two particles move off in directions making an opening angle

$$\theta + \phi = 90°.$$

Then equation **3.29** gives the momentum of the scattered particle as

$p_3 = p_1 \sin \phi = p_1 \cos \theta$.

Summarizing, we may say that the particles move off in directions making an angle 90°; each particle has momentum proportional to the cosine of the angle between its direction and that of the incident particle. In particular, this means that a recoil particle which is projected directly forwards takes all the energy of the particle which hits it. If it is projected nearly forwards, at a small angle ϕ, its energy differs from that of the incident particle by a factor $\cos^2 \phi$. So if one passes neutrons into hydrogenous material and collects protons recoiling within a cone of semi-angle 10°, their average energy is 98·5 per cent of that of the neutrons.

3.7.2 The relativistic case

Calculation. Relativistic kinematics become necessary for proton–proton scattering or neutron–proton scattering when the incident kinetic energy exceeds about 50 MeV.

The calculations are most simply done through the centre-of-mass system. We shall carry them out by putting the rest masses equal in the equations of section 3.5.5. This serves as an example in the use of the general method, but it should be pointed out that the case of equal masses also lends itself to a simplified treatment by first principles.

We start with the total energy of the incident particle in the laboratory system, which is

$E_1 = (p_1^2 c^2 + M^2 c^4)^{\frac{1}{2}}$.

Its kinetic energy T_1 is given by

$T_1 = E_1 - Mc^2$.

The velocity of the c.m.s., in units of c, is given by equation **3.19** as

$$\beta_c = \frac{p_1 c}{E_1 + Mc^2}.$$

M^*, generally known as the invariant mass of the system, is given by equation **3.20**, which simplifies to

$$M^{*2} = \frac{2M(E_1 + Mc^2)}{c^2} = \frac{2Mp_1^2}{T_1}. \qquad \textbf{3.38}$$

Equation **3.21** now gives for γ_c the expressions

$$\gamma_c = \frac{E_1 + Mc^2}{M^* c^2} = \frac{M^*}{2M} = \frac{p_1}{\sqrt{(2MT_1)}} = \sqrt{\left[\frac{E_1 + Mc^2}{2Mc^2} \right]}. \qquad \textbf{3.39}$$

The momenta of the particles, measured in the c.m.s., are obtained by reducing equation **3.22** to one of the equivalent forms

$$p' = \sqrt{\left[\frac{MT_1}{2}\right]} = \frac{M}{M^*}\,p_1 = \frac{p_1}{2\gamma_c}. \qquad \textbf{3.40}$$

E', the energy of each of the particles in the c.m.s., is given by equations **3.23** and **3.24** as

$E' = \frac{1}{2}M^*c^2$.

The components of momentum, measured in the laboratory system, are given by equation **3.25** for the scattered particle 3, which has c.m.s. angle θ_c. Equations **3.38–40** allow this to be simplified to the following more manageable pair of equations:

$p_x = \frac{1}{2}\,p_1\,(1 + \cos\theta_c),$

$p_y = \frac{p_1}{2\gamma_c}\sin\theta_c.$

The values for the recoil particle 4 are obtained by putting $\pi + \theta_c\,(= \pi - \theta_c$ on the other side) instead of θ_c. The laboratory-system angles are then given, as in equations **3.26** and **3.27**, by

$$\cot\theta = \gamma_c\,\frac{1 + \cos\theta_c}{\sin\theta_c} = \gamma_c\cot\tfrac{1}{2}\theta_c, \qquad \textbf{3.41}$$

$$\cot\phi = \gamma_c\,\frac{1 - \cos\theta_c}{\sin\theta_c} = \gamma_c\tan\tfrac{1}{2}\theta_c. \qquad \textbf{3.42}$$

Check. The equations of the preceding section reduce, in the case of small incident momentum, to the following non-relativistic equations:

$M^* = 2M + \frac{1}{2}T_1,$

$p' = \frac{1}{2}p_1,$

$E' = Mc^2 + \frac{1}{4}T_1,$

$p = \frac{1}{2}p_1(1 \pm \cos\theta_c)$ } $\left\{\begin{array}{l}+\text{for scattered particle }(\theta_c),\\ -\text{for recoil particle }(\pi+\theta_c),\end{array}\right.$

$p = \pm\frac{1}{2}p_1\sin\theta_c$

$\theta = \frac{1}{2}\theta_c,$

$\phi = \frac{1}{2}\pi - \frac{1}{2}\theta_c,$

$p_3 = p_1\cos\theta,$

$p_4 = p_1\cos\phi.$

The laboratory-system opening angle. In relativistic conditions, the opening angle $\theta + \phi$ between the directions of the two final particles is no longer equal to $\frac{1}{2}\pi$; nor is it independent of the individual angles θ and ϕ.

To obtain an idea of the magnitude of the relativistic effect, we use equation **3.41** to calculate the laboratory system angles when θ_c is 90°. Being equal, they are given by

$\cot\phi = \cot\theta = \gamma_c.$

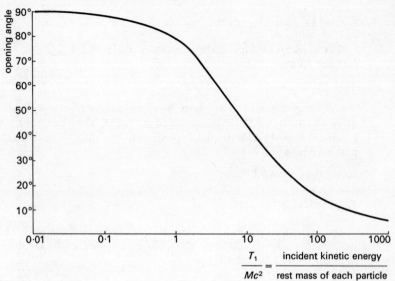

$$\frac{T_1}{Mc^2} = \frac{\text{incident kinetic energy}}{\text{rest mass of each particle}}$$

Figure 9 Elastic scattering with equal masses: opening angle for symmetrical scatters ($\theta_c = 90°$)

The opening angle $\theta + \phi$ is no longer constant, but for these symmetrical scatters it has the value $2 \cot^{-1} \gamma_c$, which decreases with increasing kinetic energy as shown in Table 5 and in Figure 9.

Table 5 Elastic Scattering with Equal Masses

T_1/MeV		$\dfrac{T_1}{Mc^2}$	γ_c	Opening angle
protons	*electrons*			$\theta + \phi$
<5	<0·002	<0·005	1·000	90·0°
9·38	0·005	0·01	1·002	89·9°
93·8	0·05	0·1	1·025	88·6°
938	0·51	1	1·225	78·4°
9380	5·1	10	2·450	44·4°
93 800	51	100	7·141	16·0°

Chapter 4
Reactions of Light Nuclei

4.1 **General points**

4.1.1 *Why light nuclei?*

In the preceding chapter we took as a standard two-body reaction the collision of a moving particle with a stationary target nucleus leading to the formation of two final particles. An energy Q was released in the process.

In practice this is a common type of reaction. For it to occur, the incident particle must approach the target nucleus closely enough for nuclear forces to take effect. In other words, the distance of closest approach between the particles must be less than the range of the nuclear forces. For a given kinetic energy of the incident particle, the distance of closest approach is proportional to the product of the charges on the two particles. Therefore reactions go most readily when these charges are small. In practice this means when the target is a light nucleus with Z less than about ten and the incident particle is a proton, a deuteron or an alpha particle. (Although important work has been done with beams of tritons and of nuclei of the helium isotope ^3_2He, it is not discussed here.)

But if the incident particle is a neutron, the charge on the target nucleus does not affect the distance of closest approach; therefore neutrons can interact with nuclei of heavy as well as of light elements. Chapter 8 is devoted to the special types of nuclear reaction caused by neutrons.

The present chapter covers the interactions of protons, deuterons and alpha particles with the nuclei of light elements, classified according to the nature of the incident and emitted particles: thus a (p, α) reaction is one in which the incident particle is a proton and the emitted one an alpha particle. After considering the general requirement of energy balance, we shall consider several types of reaction, illustrating each with examples of particular interest.

4.1.2 *Energy balance*

We have seen that, for a reaction to proceed, there must be enough energy to allow close approach of the reacting particles. This requirement is a flexible one: if there is a little less energy, approach will be a little less close, and the probability of reaction will be slightly reduced. But in sharp contrast to this smooth dependence of cross-section on energy, the requirements of the law of conservation of energy are inflexible. In the notation of Chapter 3, the product particles between them carry off kinetic energy equal to the incident kinetic energy plus any energy Q released in the reaction. Obviously if Q is negative, and the incident kinetic energy is less than $-Q$, there will be no reaction.

When a reaction can take place, the energies and momenta of the particles are related to their directions according to the kinematical equations of Chapter 3. Conversely, the same equations may be used to calculate the Q-value of a reaction from the measured values of momenta of product particles at given angles. From the Q-values of the various reactions, lists of masses of the nuclides may be built up, as indicated in section 3.2.3.

4.2 Reactions involving charged particles

4.2.1 (α, p) reactions

The earliest artificially produced nuclear reaction (Rutherford, 1919) was the disintegration of nitrogen-14 by alpha particles, according to the scheme

$$^{14}_{7}\text{N} + ^{4}_{2}\text{He} \longrightarrow ^{17}_{8}\text{O} + ^{1}_{1}\text{H} \quad (Q = -1 \cdot 198 \text{ MeV}).$$

The alpha particles were from natural radioactive decay of polonium-214; their kinetic energy of 7·7 MeV is enough to overcome the potential barrier due to the charge of the nitrogen nucleus, allowing a reasonable number of them to approach within the range of nuclear forces. $^{17}_{8}\text{O}$ is an unusual isotope of oxygen, much less stable than $^{16}_{8}\text{O}$ and the nuclei $^{14}_{7}\text{N}$ and $^{4}_{2}\text{He}$ from which it is formed in this reaction. The energy released in the reaction is therefore negative; but its magnitude is only 1·198 MeV, which is easily covered by the 7·7 MeV kinetic energy of the incident alpha particles.

Other (α, p) reactions have been observed in all the stable nuclei from $^{4}_{2}\text{He}$ to $^{19}_{9}\text{F}$, with Q-values ranging from $-17 \cdot 346$ MeV ($^{4}_{2}\text{He}$) to $+4 \cdot 071$ MeV ($^{10}_{5}\text{B}$). Most of the Q-values are negative, because the alpha particle is very stable with respect to the free proton (see section 9.1.1).

4.2.2 (p, α) reactions

The converse of an (α, p) reaction is a (p, α) reaction, in which an (artificially accelerated) proton causes a reaction with emission of an alpha particle. The Q-values of (p, α) reactions in the light nuclei are mostly positive, representing an increase of over-all stability. Exceptions are the reactions $^{12}_{6}\text{C}(\text{p}, \alpha)^{9}_{5}\text{B}$ ($Q = -7 \cdot 559$ MeV) and $^{16}_{8}\text{O}(\text{p}, \alpha)^{13}_{7}\text{N}$ ($Q = -5 \cdot 205$ MeV, but reaction not observed), in which the target nuclei are particularly stable.

In fact, the first nuclear reaction using artificially accelerated particles instead of natural alpha-particles was the (p, α) reaction of $^{7}_{3}\text{Li}$, observed by Cockcroft and Walton (1932):

$$^{7}_{3}\text{Li} + ^{1}_{1}\text{H} \longrightarrow ^{4}_{2}\text{He} + ^{4}_{2}\text{He} \quad (Q = 17 \cdot 337 \text{ MeV}).$$

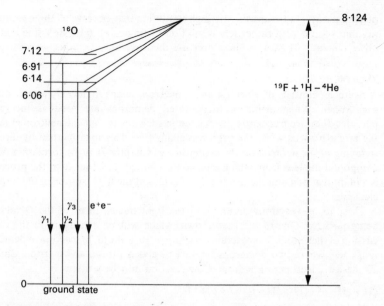

Figure 10 Energy levels of oxygen-16

It is now known that this process is accompanied by reactions in which the eight nucleons remain together as a $_4^8$Be nucleus, excited to an energy about 18 MeV above its ground state. These excited $_4^8$Be nuclei subsequently decay (with a mean life about 5×10^{-20}s) to the ground and first excited states, with emission of gamma rays of energy 17·2 and 14·8 MeV. The process as a whole may be represented by the scheme

$$_3^7\text{Li} + _1^1\text{H} \longrightarrow _4^8\text{Be} + \gamma \quad (Q = 17·242 \text{ MeV}).$$

With protons of kinetic energy 0·4–2·0 MeV, this reaction provides a convenient source of hard γ-rays. We have listed it among the (p, α) reactions because the ground state of $_4^8$Be is unstable by 0·096 MeV with respect to decay into two alpha particles, and the end results of the whole process are two alpha particles and a gamma ray resulting from transitions between levels of the intermediate nucleus $_4^8$Be.

Another well-known (p, α) reaction is used as a source of gamma rays: this is the reaction

$$_9^{19}\text{F} + _1^1\text{H} \longrightarrow _8^{16}\text{O} + _2^4\text{He} \quad (Q = 8·124 \text{ MeV}).$$

The resulting $_8^{16}$O nucleus may be formed in its ground state, with release of the whole energy Q, or it may be formed in one of four excited states at 6·06, 6·14, 6·91 and 7·12 MeV (see Figure 10). The level at 6·06 MeV decays by emitting an electron–positron pair, while the other three levels decay by emission of gamma rays. Experiments with high resolution may observe all three gamma ray lines, those with rather less resolution distinguish a 6·14 MeV line and a mixed line at 7·01 MeV, while others use the three lines together, unresolved.

4.2.3 (d, p) *reactions*

A deuteron consists of a proton and a neutron, relatively loosely attached to each other. When artificially accelerated deuterons are used as incident particles, they are frequently dissociated in such a way that the neutron enters the target nucleus while the proton remains free. This may happen by direct stripping of the neutron off the deuteron (see Chapter 7), or by formation of a compound nucleus from which a proton is emitted. In either case, the process is a (d, p) reaction and the product is a nucleus of the next isotope of the target element.

These (d, p) reactions occur in all the light nuclei, most of the Q-values being positive. One of the best-known, which will be met again in the discussion of thermonuclear reactions in Chapter 10, is the (d, p) reaction in deuterium itself, where two deuterons interact to give a proton and a triton. Since all these are isotopes of hydrogen, the reaction may be written

$$^2_1H + {}^2_1H \longrightarrow {}^1_1H + {}^3_1H \quad (Q = 3\cdot174 \text{ MeV}).$$

As an example of a (d, p) reaction in which the target nucleus is larger than the deuteron, and simply strips the neutron off it, the reaction in ^{16}O may be selected:

$$^{16}_8O + {}^2_1H \longrightarrow {}^{17}_8O + {}^1_1H \quad (Q = 1\cdot918 \text{ MeV}).$$

4.2.4 (d, α) *reactions*

When a deuteron is fully involved in a reaction, and not just dissociated by it, the range of possible reaction products includes the alpha particle. The resultant (d, α) reactions tend to have large positive Q-values, since the alpha particle is a much more tightly bound structure than the deuteron. An example is

$$^{14}_7N + {}^2_1H \longrightarrow {}^{12}_6C + {}^4_2He \quad (Q = 13\cdot570 \text{ MeV}),$$

and an example with a negative Q-value, because the product nucleus is less stable than the target, is

$$^{12}_6C + {}^2_1H \longrightarrow {}^{10}_5B + {}^4_2He \quad (Q = -1\cdot349 \text{ MeV}).$$

4.3 Reactions in which neutrons are produced
4.3.1 (α, n) *reactions*

The reactions discussed in the preceding section have involved only charged particles. Now it is necessary to consider reactions in which free neutrons are produced. The neutron was in fact discovered by Chadwick (1932) in the reaction

$$^9_4Be + {}^4_2He \longrightarrow {}^1_0n + {}^{12}_6C \quad (Q = 5\cdot708 \text{ MeV}).$$

Alpha particles from polonium were allowed to fall on a foil of beryllium metal and the neutrons were detected by the recoil protons which they knocked forward in paraffin wax.

Of the neutron sources at present on the market, a very large number use this reaction in powdered beryllium mixed with polonium-210, americium-241 or some other source of alpha particles. Other neutron sources use the (α, n) reactions in $^{11}_{5}B$, $^{19}_{9}F$ or $^{7}_{3}Li$.

The maximum energy of neutrons from any of these (α, n) reactions may be calculated by means of the kinematical equations of Chapter 3, using the Q-value of the reaction and the kinetic energy of the alpha particles from the particular radioactive source. In fact, however, neutrons will be emitted with a spectrum of energies extending from zero up to this maximum, because of two effects: first the direction of emission of the neutron is not defined; second, the alpha particles are slowed down by passage through the material of the source or the target if they do not happen to come outwards from a nucleus on the surface of a source grain and interact with the first target nucleus that they approach. Thus each type of source has a characteristic spectrum of neutron energies, which may determine its appropriateness for a particular purpose.

4.3.2   (d, n) *reactions*

It has already been mentioned that an incident deuteron may be split up in a reaction, the neutron from it remaining in the target nucleus so that the result is a (d, p) reaction. On the other hand, it may be the proton that remains in the target nucleus, leaving the neutron free; in this case the result is a (d, n) reaction.

Deuterons can in fact produce neutrons when incident upon almost any target material. This fact, together with the possibility of the reaction $^{2}H(d, n)^{3}H$ (see below) occurring with deuterons already trapped in the target, imposes severe limitations on the extent to which apparatus can be prevented from producing neutrons if any contamination remains from previous use with deuterium.

As an example of a (d, n) reaction in a light nucleus, in which the neutron spectrum is determined by the properties of the product nucleus, we may quote that in the isotope $^{7}_{3}Li$,

$$^{7}_{3}Li + ^{2}_{1}H \longrightarrow ^{8}_{4}Be + ^{1}_{0}n \quad (Q = 15 \cdot 017 \text{ MeV}).$$

The exact energies of the neutrons depend upon the angle of observation, and on the kinetic energy of the incident deuterons, but a typical spectrum is shown in Figure 11, where the peak due to formation of beryllium-8 in its ground state is distinguishable above the spread which results from formation of beryllium-8 in excited states.

An important (d, n) reaction, which gives only a single group of neutrons, is the reaction of two deuterons, which has already been mentioned,

$$^{2}_{1}H + ^{2}_{1}H \longrightarrow ^{1}_{0}n + ^{3}_{2}He \quad (Q = 3 \cdot 268 \text{ MeV}).$$

55   Reactions in which neutrons are produced

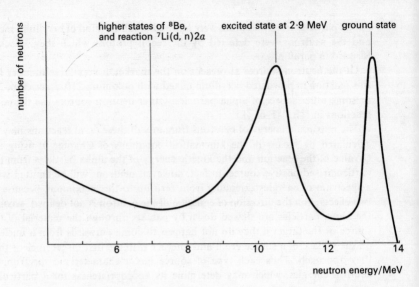

Figure 11 Energy spectrum of neutrons from the reaction ^{7}Li(d, n)^{8}Be at 120°; deuteron energy 0·93 MeV

The energy of this group of neutrons is a function of the angle of observation, and of the kinetic energy of the incident deuterons. The student may check, as an exercise in kinematics, that for a given incident energy there exists an angle of observation at which the neutron energy is to a first order unaffected by change of incident energy. For example, a nearly homogeneous group of neutrons of energy 2·4 MeV is obtained at 107° when a 1 MeV deuteron beam falls upon a target containing deuterium, even if the deuterium is distributed over a target thickness sufficient to slow down or even stop the incident deuterons. (For experimental confirmation of this, and of the angular dependence of neutron energy, see Livesey and Wilkinson, 1948).

As a third example of a (d, n) reaction, the choice falls on one which has become increasingly important since tritium, the third isotope of hydrogen, became available in sufficient quantities to form targets,

$$^2_1\text{H} + ^3_1\text{H} \longrightarrow ^1_0\text{n} + ^4_2\text{He} \quad (Q = 17\cdot577 \text{ MeV}).$$

Since the Coulomb barrier is a minimum for reactions between isotopes of hydrogen, this reaction can be made to occur with deuterons which have been accelerated through only a few kilovolts. In fact, a resonance in the compound nucleus ^{5}He causes the cross-section to have a peak at 107 keV. The relatively large Q-value causes the neutrons to be produced with energies around 14·3 MeV, with an angular dependence determined by the incident energy. A common type of commercially available neutron generator (see section 8.2.2) consists of a sealed tube containing a target of a rare-earth tritide, and gaseous deuterium from which ions can be accelerated onto the target. Alternatively, a deuteride may be used as the target, with gaseous tritium to provide the accelerated ions. In practice the target life may be increased by using the two systems mixed, if the presence of some neutrons from the reaction $^2H(d, n)^3He$ does not matter.

4.3.3 (γ, n) reactions

Photoneutron or (γ, n) reactions, may be made to occur in many target elements. The (γ, n) reaction in beryllium has $Q = -1.666$ MeV, which allows its use with γ-rays from radioactive elements such as antimony-124, radium-226 and thorium-228 in neutron sources, for special purposes.

An interesting (γ, n) reaction is the photodisintegration of deuterium. This is just the break-up of a deuteron into its constituent neutron and proton. The gamma ray must have energy greater than the binding energy of the deuteron (2·225 MeV), any excess becoming kinetic energy of the emitted neutron and proton. This reaction occurs simply, without competing processes, with gamma rays of energy up to about 150 MeV, above which meson production starts to occur.

4.4 Excited states of nuclei

It has been mentioned, apropos of several of the reactions described above, that the product nucleus can be formed either in the ground state or in an excited state. The study of these excited states constitutes a form of nuclear spectroscopy. Catalogues of the known excited states of light nuclei, giving a good indication of the volume of work which has been done in this field, are provided by Ajzenberg–Selove and Lauritsen (1959, 1966, and 1968).

The first experimental step is to prove the existence of a level at a particular energy, from observations on the products of one or more reactions in which it is formed. Then the quantum numbers of the level must be found: its angular momentum and parity may in some cases be deduced from the angular distributions of reaction products, or from their angular correlation. A further quantum number, important in some cases, is the isospin, which is discussed in Chapter 11.

Nuclear levels tend to be irregularly spaced and do not lend themselves to any numerology as simple as that which led to the interpretation of atomic spectra. Many of the excited levels have, however, found satisfactory interpretation, along with the ground states, in terms of the shell model of the nucleus. This is discussed by Reid (1971).

Chapter 5
Wave-Mechanical Theory of Scattering

5.1 Cross-sections

5.1.1 *Probability*

In Chapter 3 we discussed the kinematic conditions which must be obeyed by a nuclear reaction, if it happens at all. But we said nothing about the probability that the reaction would take place, nor about the relative probabilities that the products would be emitted in particular directions. Some directions were shown to be kinematically impossible, but there was no discussion about which of the possible directions would be observed most frequently.

In the present chapter, we develop a formalism for describing the probability of a nuclear process, in terms of a wave-mechanical transition amplitude. We give particular attention to the especially simple process of elastic scattering, where the description naturally covers the angular distribution of the products, without the complication of extra parameters to describe new types of product particle.

5.1.2 *Definition of cross-section*

To describe the probability of a particular nuclear process, or the frequency with which it occurs when a large number of particles have the opportunity of undergoing it, we imagine target particles spread randomly over a plane, with a uniform density n per unit area. The incident particles are allowed to fall perpendicularly on this plane, as a uniform parallel beam.

The probability that a given incident particle will undergo the specified reaction is clearly proportional to n, the number of target particles per unit area. If we make the probability equal to σn, we may say that each target particle is presenting an effective cross-sectional area σ. We say that the cross-section for the process is σ, meaning that a fraction σn of the particles in an incident beam will undergo it. The cross-section σ is not normally related to any geometrical area. Indeed, a given type of target nucleus may have a cross-section σ_1 for one process, σ_2 for a second process, σ_3 for a third, and so on; in this case, the total number of particles removed from the beam will be given by the total cross-section σ_{tot}, which is the sum of the cross-sections for all

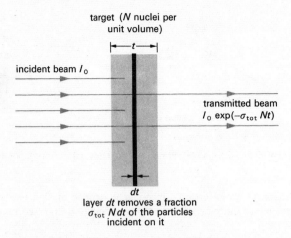

incident beam I_0

transmitted beam $I_0 \exp(-\sigma_{tot} Nt)$

dt

layer dt removes a fraction $\sigma_{tot} N\, dt$ of the particles incident on it

Figure 12 Absorption of particles in beam

possible processes of interaction (elastic scattering, inelastic scattering, capture, reactions, etc.).

It should be noted that this simple proportionality holds also in a real target, which is not a plane but a slab of non-zero thickness, so long as $\sigma_{tot} n \ll 1$. If this condition does not hold, the beam of incident particles is attenuated by the front layer of target nuclei, and fewer interactions occur in the back layers. To handle this situation, we consider a layer of infinitesimal thickness dt, containing $N\, dt$ target nuclei per unit area, where N is the number of target nuclei per unit volume. This layer will remove a fraction $\sigma_{tot} N\, dt$ of the particles incident upon it. The incident beam is therefore absorbed exponentially, its intensity falling off with target thickness t according to $\exp(-\sigma_{tot} Nt)$. The total number of interactions in a layer of thickness t is therefore (see Figure 12)

$$\{1 - \exp(-\sigma_{tot} Nt)\} \times (\text{number of incident particles}).$$

The total cross-section is sometimes called the absorption cross-section, since it gives the rate of absorption of particles from the beam. Indeed it is often measured by comparing the intensities of a beam before and after passing through a target foil. However, we shall call it the total cross-section, since absorption can be used as a synonym for capture.

5.1.3 Differential cross-section

Just as the probability for a process may be divided into partial probabilities, the cross-section may be divided up according to the angle of emission θ of one of the final particles. We may consider the part of the process leading to emission into a chosen element of solid angle $\delta\Omega$ as a process in itself, having cross-section $\delta\sigma$. The cross-section per unit solid angle, written $d\sigma/d\Omega$, is

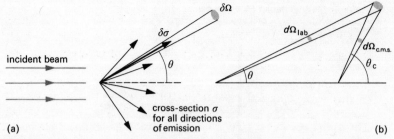

Figure 13 Differential cross-section. (a) Principle; (b) conversion from laboratory to c.m.s.

known as the differential cross-section. Since it may depend on θ, it needs to be specified, either numerically or analytically, as a function of θ (see Figure 13a).

The angle θ and the solid angle $\delta\Omega$ may be specified either in the laboratory system or in the centre-of-mass system. To transform a differential cross-section from the laboratory system to the centre-of-mass system, it must be multiplied by the ratio $d\Omega_{\text{lab}}/d\Omega_{\text{c.m.s.}}$. This ratio may be calculated from the kinematics of the reaction, by way of the relation between the angles in the two systems; it is a function of the angle of observation, and of the energy of the incident particles (see Figure 13b).

The simplest process is one which is isotropic in the centre-of-mass system. For such a process the differential cross-section is independent of $\theta_{\text{c.m.s.}}$ and the cross-section σ is equal to $4\pi(d\sigma/d\Omega)_{\text{c.m.s.}}$.

5.1.4 The unit of cross-section

Cross-sections are usually measured in terms of a unit of 10^{-24} cm^2, colloquially known as a barn. Smaller cross-sections may be expressed in millibarns (10^{-27} cm^2) or microbarns (10^{-30} cm^2). A differential cross-section might be expressed as so many millibarns per steradian.

In a real target of thickness t, the fraction of the beam interacting is

$$\sigma N = \sigma \frac{L}{A} t\rho,$$

where L is Avogadro's constant ($6{\cdot}023 \times 10^{23}$ mol^{-1}), A is (mass number of target nuclei) $\sim$ (atomic weight of target atoms), and ρ is the density. For example, a cross-section of one millibarn, in a carbon target of density 2, 1 mm thick, would cause about 10^{-5} of the beam to interact.

5.2 Semi-classical model of scattering

5.2.1 *Mechanical model*

In a crudely mechanical model, obeying classical laws, we might imagine a beam of small pellets, falling on a circular target constructed out of boxes and tilted plates so that it captured some incident particles but scattered others (elastically in so far as the construction was ideal). It might have effective cross-sections σ_{cap} and σ_{scat} for these two processes. The sum of these two cross-sections would be σ_{tot}, the total cross-section; this would be equal to the geometrical cross-section πR^2, unless there were holes through which particles could pass without suffering either capture or scattering. By suitable choice of method of construction, we could obtain any desired values of σ_{cap} and σ_{scat}. Apart from the limit on their sum, the values could be chosen independently. The unaffected part of the beam would continue past the target, throwing a sharply defined shadow.

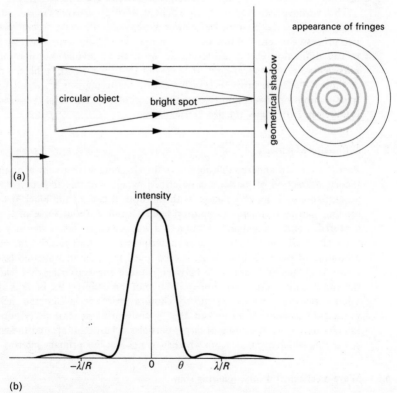

Figure 14 Diffraction of light by circular obstacle. (a) Fresnel diffraction, showing bright spot on axis. (b) Distribution of intensity as a function of angle of scattering (R = radius of obstacle)

5.2.2 Diffraction of light

When our incident beam is of light, we know that the mechanical model is too crude: the shadow cannot be perfectly sharp, because even a perfectly black disc produces a diffraction pattern (see Figure 14a). All the light falling on the disc is captured, but some of the light which passes it changes direction. There is a bright spot at the centre of the shadow, where light from all points on the rim of the disc arrives in the same phase. Inside the shadow, and extending beyond it, there are alternating bright and dark bands, with angular spacing of order λ/R, where λ is the wavelength of the light and R the radius of the disc. In the process of absorbing part of the incident beam, the disc causes a redistribution of the energy of the waves which are not absorbed; instead of continuing in their original direction outside the geometrical shadow, some of them change direction and move either into or away from the shadow. This change of direction may be called elastic scattering of photons. A typical distribution of intensity with direction is shown in Figure 14(b).

Thus we have seen that, in the case of light, absorption cannot occur without being accompanied by some diffraction. We describe the relation between the two processes in terms of the wave-like properties of light; only when we want to call diffraction elastic scattering do we need to invoke the quantum or particle-like properties of light. It turns out that for a perfectly black sphere of radius $R \gg \lambda$, the cross-section for elastic scattering is πR^2, equal to the cross-section for absorption (i.e. capture of photons). The total cross-section is therefore $2\pi R^2$, twice the geometrical area.

5.2.3 Diffraction of particles

Particles too can undergo diffraction. In the elementary treatment of quantum theory, diffraction of particles is described in terms of the de Broglie waves accompanying them. By analogy with the optical case of the black disc, we see that particles cannot be captured from a beam without some of the uncaptured particles changing direction in a way which makes us say they have been elastically scattered. Effective descriptions of the capture and elastic scattering of particles must therefore be given together in wave-mechanical terms. It is the close connexion between capture and scattering that leads to the appearance of the 'scattering' in the title of this chapter; the book is about nuclear reactions, which are mostly capture processes in that the incident particles disappear. It turns out that a convenient general description of interactions between beam and target particles is best built around the elastic scattering, which always occurs whether or not it is our primary interest.

5.3 Wave-mechanical theory, ignoring spin

5.3.1 Plane waves

Let us describe our beam of incident particles, advancing along the z-axis, as a set of plane de Broglie waves. The wave function in this plane wave is

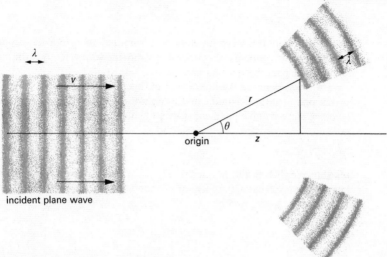

Figure 15 Scattering by target at the origin

$\psi_{\text{inc}} = e^{ikz}e^{-i\omega t}$.

In this expression, t represents time, ω is the angular frequency, given in terms of the total energy E of each incident particle by

$$\omega = \frac{E}{\hbar}.$$

k is the wave number, given by

$$k = \frac{1}{\lambdabar} = \frac{2\pi}{\lambda} = \frac{p}{\hbar},$$

where p is the momentum, λ the wavelength, and h Planck's constant. The bars through $\lambdabar$ and $\hbar$ indicate insertion of a factor $1/2\pi$, as usual.

In quantum-mechanical arguments it is common to use a simplified wave function which does not include the oscillatory term $e^{-i\omega t}$. This device makes calculation simpler, but it may be necessary at some stages to insert the omitted term to make plain the physical significance. Following this idea, we use for our incident wave a simplified wave function

$$\psi_{\text{inc}} = e^{ikz}.$$

This may be rewritten in the notation of Figure 15 as

$$\psi_{\text{inc}} = e^{ikr\cos\theta}.$$

Our plane wave has

$$|\psi_{\text{inc}}|^2 = 1.$$

This means that the wave function is normalized to give a density of one particle per unit volume. If the particles are moving with velocity v, the incident flux will be v particles per unit area per second.

We want to describe the interaction of this plane wave with a single target particle placed at the origin. To do this we use a mathematical device to replace the plane wave by an equivalent system of coherent spherical waves.

5.3.2   *Spherical waves*

It is shown by Mott and Massey (1933, p. 22) that a plane wave may be considered as a coherent superposition of spherical waves, which at large r have an asymptotic form given by

$$\psi_{\text{inc}} = e^{ikr \cos \theta} = \frac{1}{2ikr} \sum_l (2l+1)i^l P_l(\cos \theta)\{e^{i(kr - \frac{1}{2}l\pi)} - e^{-i(kr - \frac{1}{2}l\pi)}\}. \qquad \textbf{5.1}$$

The two exponential terms in the curly bracket are obtained from a single term $\sin(kr - \frac{1}{2}l\pi)$. In their exponential form, we can see that the first of them represents an outgoing spherical wave, because when we include the oscillatory factor $e^{-i\omega t}$ to reconstitute the time-dependent wave function ψ_{inc} we get $e^{i(kr - \omega t)}$, which keeps a given value if r increases as t increases. It therefore represents an outgoing spherical wave, while the second term represents an ingoing wave shrinking towards the origin.

The spherical waves are called partial waves. Each is characterized by an orbital angular momentum quantum number l. It is spherical in that it moves at a rate depending on r, not on θ; but its amplitude is not in general isotropic, having an angular distribution given by the Legendre polynomial $P_l(\cos \theta)$. These polynomials are discussed in many books, and a general expression for the nth polynomial is given by Menzel (1960). The first few are as follows:

$l = 0: P_0(\cos \theta) = 1,$

$l = 1: P_1(\cos \theta) = \cos \theta,$

$l = 2: P_2(\cos \theta) = \frac{1}{2}(3 \cos^2\theta - 1),$

$l = 3: P_3(\cos \theta) = \frac{1}{2}(5 \cos^3\theta - 3 \cos \theta),$

$l = 4: P_4(\cos \theta) = \frac{1}{8}(35 \cos^4\theta - 30 \cos^2\theta + 3).$

$P_0(\cos \theta)$ is isotropic, while higher polynomials contain powers of $\cos \theta$ up to the lth.

5.3.3 *The effect of a target particle*

Scattering potential. Let us consider a spinless target particle, placed at the origin. It will have a spherically symmetric potential, whose effects will be important over a limited region around it. The effect of the target particle on the incident waves may be described as a modification of some or all of the outgoing partial waves, as they move outwards through the potential. The ingoing spherical waves cannot be affected because, in common language, they have not yet arrived at the place where they could feel the influence of the target particle. In more high-brow language, one says that 'the ingoing waves cannot be affected because of causality'.

The need for the centre-of-mass system. If the target particle is not infinitely heavy, we must put the centre-of-mass at the origin, and describe what happens in the centre-of-mass system of coordinates. This device is necessary, to ensure that incident particles and particles scattered in different directions all have the same momentum. They can thus be described in terms of spherical waves with a single value of the wave number k. In the laboratory system, elastically scattered particles have momenta depending on direction, and each direction would need its own k.

The phase shift and elasticity. Let us suppose that the lth outgoing partial wave is modified to an extent described by a complex multiplying factor $\eta_l e^{2i\delta_l}$. The angle δ_l is a real angle, known as the phase shift, while η_l is a real number called the elasticity, for reasons which will become apparent later.

The total wave function, describing the whole system of waves, is now

$$\psi_{\text{tot}} = \frac{1}{2ikr} \sum_l (2l+1)i^l\, P_l(\cos\theta)\, \{\eta_l\, e^{2i\delta_l}e^{i(kr-\frac{1}{2}l\pi)} - e^{-i(kr-\frac{1}{2}l\pi)}\}. \qquad \textbf{5.2}$$

Subtracting the wave function of the incident beam **5.1**, we obtain a wave function which must represent the elastically scattered particles,

$$\dot\psi_{\text{scat}} = \frac{1}{2ikr} \sum_l (2l+1)\, P_l(\cos\theta)\, (\eta_l\, e^{2i\delta_l} - 1)\, e^{ikr}. \qquad \textbf{5.3}$$

This expression is simpler than the functions **5.1** and **5.2** because, when the ingoing spherical waves disappear in the subtraction, the factor i^l is cancelled by the factor $e^{-i\frac{1}{2}l\pi}$ in the outgoing wave. From the fact that this expression includes only the single wave number k we can confirm that it represents elastically scattered particles, and that the whole argument applies to the centre-of-mass system, as indicated in the previous section.

The transition amplitude. The wave function of the scattered particles may be expressed in the alternative form

$$\psi_{\text{scat}} = \frac{e^{ikr}}{kr} \sum_l (2l+1)\, P_l(\cos\theta)\, T_l. \qquad \textbf{5.4}$$

T_l is called the transition amplitude, since it describes the new part of the lth partial wave, created by transitions from the same partial wave in its original form. To make expression **5.4** equivalent to **5.3**, the transition amplitude has to be a complex number given by

$$T_l = \frac{\eta_l\, e^{2i\delta_l} - 1}{2i}.$$ **5.5**

The relations between η_l, δ_l and T_l are shown plotted in the complex plane in Figure 16. Diagrams of this type are often used to plot the way in which elastic scattering in a particular partial wave varies with incident energy. This is especially useful for resonance scattering (see Chapter 6).

The scattering amplitude. A third alternative expression for ψ_{scat} takes the much simpler form

$$\psi_{\text{scat}} = \frac{e^{ikr}}{r} f(\theta).$$

The function $f(\theta)$ is called the scattering amplitude, and is given by

$$f(\theta) = \lambda \sum_l (2l+1)\, P_l(\cos\theta)\, T_l.$$ **5.6**

(This looks tidier with λ written for k^{-1}.)

5.3.4   *The differential cross-section for elastic scattering*

The general expression. We shall now obtain an expression for the differential cross-section for elastic scattering, in terms of the scattering amplitude.

The density of particles in the scattered wave, at a point defined by co-ordinates (r, θ), is given by

$$|\psi_{\text{scat}}|^2 = r^{-2}|f(\theta)|^2.$$

Since they are elastically scattered particles, described by a wave function containing the same wave number k as was used for the incident particles, they will be moving outward from the origin with the velocity of the incident particles. The flux of scattered particles, per unit area per second, is therefore

$$vr^{-2}|f(\theta)|^2.$$

Unit solid angle at distance r fills an area r^2; therefore the flux of scattered particles per second into an element $d\Omega$ of solid angle is

$$v|f(\theta)|^2 d\Omega.$$

Since the incident beam contains v particles per unit area per second, the differential cross-section, defined as number of scattered particles per unit solid angle, for one incident particle per unit area is

$$\frac{d\sigma}{d\Omega} = |f(\theta)|^2.$$ **5.7**

The case of only one partial wave. If only one partial wave is important, the summation in $|f(\theta)|^2$ leads to no complications, and we may write

$$\frac{d\sigma}{d\Omega} = |f(\theta)|^2 = \lambda^2(2l+1)^2 \, [P_l(\cos\theta)]^2 \, |T_l|^2.$$

The elastically scattered particles then have an angular distribution given by the square of the *l*th Legendre polynomial. This means that in the case of pure S-wave scattering, with $l = 0$, the scattering is isotropic. On the other hand, predominantly P-wave scattering, with $l = 1$, gives an angular distribution proportional to $\cos^2\theta$.

The case of many partial waves – interference. When more than one partial wave is present, the summation gives rise to cross-terms as well as to the terms due to the partial waves separately. The cross-terms may be described physically as resulting from interference between the pairs of partial waves. To get the general expression for $d\sigma/d\Omega$, we have to take a sum over l in $f(\theta)$, and take a separate sum over a different l' in the complex conjugate $f^*(\theta)$, as follows:

$$\frac{d\sigma}{d\Omega} = |f(\theta)|^2$$

$$= f(\theta)f^*(\theta)$$

$$= \left[\sum_l (2l+1) \, P_l(\cos\theta) \, T_l\right] \left[\sum_{l'} (2l'+1) \, P_{l'}(\cos\theta) \, T_{l'}^*\right]$$

$$= \sum_l (2l+1)^2 \, [P_l(\cos\theta)]^2 \, |T_l|^2 \, +$$

$$+ \sum_{l,\,l'} (2l+1)(2l'+1) \, P_l(\cos\theta) \, P_{l'}(\cos\theta) \, T_l \, T_{l'} \quad (l \neq l'). \tag{5.8}$$

The case of two partial waves. If there are only two important partial waves, equation **5.8** contains only two principal terms and one interference term. The commonest case of this type would be S-wave and P-wave scattering, with $l = 0$ and 1. For this case, the differential cross-section is

$$\frac{d\sigma}{d\Omega} = |T_0|^2 + 9(\cos^2\theta)\,|T_1|^2 + 3(\cos\theta)\,(T_0\,T_1^* + T_1\,T_0^*)$$

$$= |T_0|^2 + 9(\cos^2\theta)\,|T_1|^2 + 6(\cos\theta)\,\mathrm{Re}\,T_0\,T_1. \tag{5.9}$$

Here * indicates complex conjugate, and Re means the real part of whatever follows.

Equation **5.9** is a particularly interesting expression: the principal terms, one isotropic and one anisotropic, both have backward–forward symmetry, since $\cos^2\theta$ is the same as $\cos^2(\pi-\theta)$. But if the interference term is positive in the forward hemisphere, it is negative in the backward one. This type of forward–backward asymmetry is liable to occur whenever partial waves of even l interfere with partial waves of odd l. In scattering processes this phenomenon implies nothing peculiar, but when it occurs in decay processes

(β-decay, or decays of mesons or hyperons), the presence of even and odd partial waves together implies non-conservation of parity (see Hughes, 1971).

5.3.5 *The over-all cross-section for elastic scattering*

The cross-section itself may be obtained by integrating the differential cross-section over the whole solid angle of 4π steradians. For this we need to use two special properties of Legendre polynomials, namely:

(a) They are all normalized, so that

$$\int_{4\pi} [P_l(\cos\theta)]^2 d\Omega = \frac{4\pi}{(2l+1)}.$$

(b) They are orthogonal, which means that cross-products disappear when integrated,

$$\int_{4\pi} P_l(\cos\theta)\, P_{l'}(\cos\theta) = 0 \quad \text{if } l \neq l'.$$

The interference terms in the expressions for the differential cross-section therefore make no contribution to the over-all cross-section, which is:

$$
\begin{aligned}
\sigma_{\text{elastic}} &= \int_{4\pi} |f(\theta)|^2 \, d\Omega \\
&= \lambda^2 \sum_l (2l+1)^2 \, |T_l|^2 \int_{4\pi} [P_l(\cos\theta)]^2 \, d\Omega \\
&= 4\pi\, \lambda^2 \sum_l (2l+1)\, |T_l|^2.
\end{aligned}
\qquad \textbf{5.10}
$$

This is a much simpler expression than that for $d\sigma/d\Omega$, because of the absence of interference terms. From it we deduce that, as in optical interference, these terms simply shift intensity from one part of the angular distribution to another, without changing the total amount of scattering.

5.3.6 *Inelastic processes*

We must now see whether our target particle is causing any loss of particles from the system which has wave number k.

In the incident wave we may consider that the ingoing and outgoing partial waves, since they have equal amplitudes, are carrying equal numbers of particles into and away from the origin. There is thus neither loss nor creation of particles at the origin. But when there is a target particle at the origin (or at a point such that the centre of mass is at the origin), the outgoing spherical waves have amplitudes changed by the factor η_l. Using the two terms in equation **5.2** separately, we see that the outgoing partial waves represent a total outward flow of particles at a rate

$$v \int\limits_{4\pi} r^2 \, d\Omega \left| \frac{1}{2kr} \sum_l (2l+1) \, P_l(\cos\theta) \eta_l \, e^{2i\delta_l} e^{ikr} \right|^2 = v \frac{1}{4k^2} \sum_l (2l+1) \, 4\pi \cdot \eta_l^2$$

$$= v \cdot \pi\lambda^2 \sum_l (2l+1) \, \eta_l^2.$$

Correspondingly, the ingoing partial waves represent a flow of particles towards the origin at a rate

$$v \cdot \pi\lambda^2 \sum_l (2l+1).$$

Subtracting these, we see that the whole system of waves involves a net disappearance of particles at the origin, at a rate

$$v \cdot \pi\lambda^2 \sum_l (2l+1) \, (1-\eta_l^2).$$

Comparing this with the incident flux of v particles per square centimetre per second, we obtain a cross-section for disappearance of particles

$$\sigma_{\text{inelastic}} = \pi\lambda^2 \sum_l (2l+1) \, (1-\eta_l^2). \qquad \textbf{5.11}$$

We call this $\sigma_{\text{inelastic}}$ because it includes all processes other than elastic scattering: nuclear reactions, capture, and all types of inelastic process. We have said nothing about the products of these processes; to describe them would need an altogether separate system of waves, each with its own wave number. We have simply calculated the rate of disappearance of particles from the system of wave number k.

It can now be seen why η_l is called the elasticity; equation **5.11** shows that when $\eta_l = 1$, the scattering is perfectly elastic and $\sigma_{\text{inelastic}}$ is zero. On the

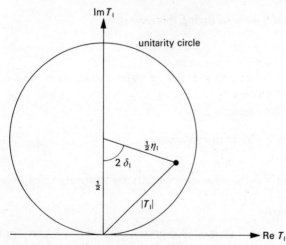

Figure 16 Transition amplitude T_l plotted in the complex plane

other hand, $\eta_l = 0$ indicates that the inelasticity is as large as it can be, the outgoing spherical wave being completely suppressed. But we must note that this does not mean absence of elastic scattering (see equations **5.10** and **5.5**).

The value of η_l is limited to between 0 and 1: a value greater than 1 would imply the appearance of more particles in the elastically scattered wave than were present in the incident wave. This is forbidden by common sense, embodied for this purpose in the principle of unitarity. For this reason, we give the name 'unitarity circle' to the circle of radius one half in Figure 16. Purely elastic scattering is represented by values of T_l lying on this circle, and all other possible values of T_l lie inside it.

5.3.7 Phase shifts

The phase shift δ_l has a definite meaning in the general case: $2\delta_l$ is the angle by which the target alters the phase of the outgoing partial wave.

In the case of purely elastic scattering, δ_l takes on a more direct significance as the phase of the scattered wave. The following argument shows that in this case the cross-section and the differential cross-section take on especially simple forms.

When $\eta_l = 1$ the transition amplitude becomes

$$T_l = \frac{e^{2i\delta_l} - 1}{2i} = e^{i\delta_l} \sin \delta_l \qquad\qquad 5.12$$

and $|T_l|^2 = \sin^2 \delta_l$.

The differential cross-section for purely elastic scattering is then

$$\frac{d\sigma}{d\Omega} = \sum_l [\lambda(2l+1) P_l (\cos \theta) \sin \delta_l]^2.$$

In the case of pure S-wave scattering, this reduces to

$$\frac{d\sigma}{d\Omega} = (\lambda \sin \delta_0)^2. \qquad\qquad 5.13$$

The over-all cross-section, with one or more partial waves, and no inelastic processes occurring, is

$$\sigma_{\text{elastic}} = 4\pi\lambda^2 \sum_l (2l+1) \sin^2 \delta_l.$$

For S-wave scattering alone this reduces to

$$\sigma_{\text{elastic}} = 4\pi\lambda^2 \sin^2 \delta_0, \qquad\qquad 5.14$$

which we can check is the same as $4\pi (d\sigma/d\Omega)$, the scattering being isotropic.

5.3.8 The total cross-section

The total cross-section, covering all processes which remove particles from the incident beam, is obtained by adding equations **5.10** and **5.11** for σ_{elastic} and $\sigma_{\text{inelastic}}$ respectively:

$$\sigma_{total} = \sigma_{elastic} + \sigma_{inelastic}$$

$$= \pi \lambda^2 \sum_l (2l+1) \{4 \, | \, T_l \, |^2 + (1 - \eta_l^2)\}.$$

The expression in the curly bracket may be simplified by means of the relation **5.5** between T_l and η_l. It is in fact just $4 \, \text{Im} \, T_l$, where $\text{Im} \, T_l$ means the imaginary part of the transition amplitude.

The total cross-section may therefore be written as

$$\sigma_{total} = 4\pi \, \lambda^2 \sum_l (2l+1) \text{Im} \, T_l. \qquad \textbf{5.15}$$

We now begin to see the direct physical significance of the transition amplitude T_l. Its imaginary part gives the total cross-section for removal of particles from the incident beam, while its modulus squared gives the cross-section for elastic scattering.

5.3.9 *Maximum values*

Many factors may conspire to reduce the value of a cross-section. For example, there may be a potential barrier to be overcome, or there may be other competing processes.

But if we concentrate on a single partial wave, and remember that $| \, T_l \, |^2$ cannot be greater than 1, we see from equation **5.10** that the cross-section for elastic scattering cannot exceed a maximum value

$$\sigma_{elastic(max)} = 4\pi \lambda^2 (2l+1). \qquad \textbf{5.16}$$

This limit may be reached if the scattering is perfectly elastic and the phase shift δ_l is 90°.

The cross-section for inelastic processes, however, reaches its maximum value when $\eta_l = 0$. Equation **5.11** shows that the upper limit is

$$\sigma_{inelastic(max)} = \pi \lambda^2 (2l+1). \qquad \textbf{5.17}$$

Since the maximum values of $\sigma_{elastic}$ and $\sigma_{inelastic}$ are obtained with different values of η_l, the maximum value of σ_{total} is not obtained by adding expression **5.16** to **5.17**. On the contrary, it is obtained by putting $\eta_l = 1$, $\delta_l = 90°$, into equation **5.15**, which gives

$$\sigma_{total(max)} = 4\pi \lambda^2 (2l+1).$$

This is the same as the upper limit for $\sigma_{elastic}$, and is obtained when the scattering is purely elastic. The dependence of elastic, inelastic and total cross-sections on η_l, for $\delta_l = 90°$, is shown in Figure 17.

We may summarize the conclusions by saying that the total cross-section has a maximum when there are no inelastic processes, giving

$$\sigma_{total(max)} = \sigma_{elastic(max)} = 4\pi \lambda^2 (2l+1). \qquad \textbf{5.18}$$

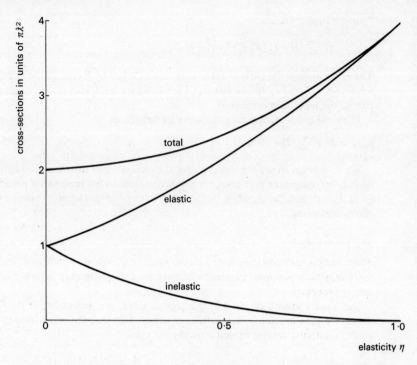

Figure 17 Elastic, inelastic and total cross-sections as a function of elasticity η for a single partial wave ($l = 0$) at resonance ($\delta = 90°$)

But the maximum value of $\sigma_{\text{inelastic}}$ is obtained in conditions which make

$$\sigma_{\text{inelastic(max)}} = \sigma_{\text{elastic}} = \pi \lambda^2 (2l+1) \qquad \textbf{5.19}$$

and

$$\sigma_{\text{total}} = \pi \lambda^2 (2l+1) = \tfrac{1}{2}\sigma_{\text{total(max)}}.$$

5.4 How many partial waves?

In a given process, how many partial waves do we need to consider? The answer, in all cases where the method of partial waves is to be useful, is 'the first few'. A quantative explanation of why, and what determines whether this means 1, 2, 3 or more, is given on two alternative levels in sections 5.4.1 and 5.4.2.

5.4.1 *The centrifugal barrier*

To describe the motion of an incident particle in the field of a target particle, we use the Schrödinger equation in the form

$$\nabla^2 \psi + \frac{2m}{\hbar^2}(E - V)\psi = 0.$$ 5.20

Here m and E are the mass and energy of the incident particle, $V = V(r)$ is the potential due to the target, assumed spherically symmetric. To obtain the radial dependence of ψ separate from its angular dependence, we express it as the product of a radial wave function $R(r)$ and a spherical harmonic $Y_l^m(\theta, \phi)$:

$$\psi = R(r)Y_l^m(\theta, \phi).$$ 5.21

The suffix l in the spherical harmonic is the angular-momentum quantum number, and m is its z-component. $R(r)$ is then a solution of the equation

$$\frac{d^2}{dr^2}\{r\,R(r)\} + \frac{2m}{\hbar^2}\left[E - V(r) - \frac{l(l+1)\hbar^2}{2mr^2}\right]r\,R(r) = 0,$$ 5.22

This looks like the one-dimensional Schrödinger equation, with $r\,R(r)$ as variable, and with potential given by the actual potential $V(r)$ plus an extra term $l(l+1)\hbar^2/2mr^2$. In a region beyond the range of nuclear interactions with the target, $V(r)$ will be zero, but the extra term will be positive. If there is a region in which $V = 0$ and

$$E < \frac{l(l+1)\hbar^2}{2mr^2},$$ 5.23

the incident particle cannot reach the target, except by quantum-mechanical tunnelling through the region, which behaves like a potential barrier (see Figure 18, and Chapter 10). There is thus a strong tendency for interactions to occur only in partial waves with l less than the limit set by inequality 5.23. By analogy with the Coulomb potential barrier, we may say that interaction in a higher partial wave requires penetration of a potential barrier.

The inequality 5.23 may be rewritten as an equation for the upper limit to l, in terms of the incident momentum $p = \sqrt{(2mE)}$ and ρ, the radius of the target nucleus (the potential radius, which includes the range of the nuclear forces),

$$\sqrt{\{l(l+1)\}} = \frac{p\rho}{\hbar} = \frac{\rho}{\lambda}.$$ 5.24

At this point, we must note that the above discussion applied to interactions through the nuclear potential. $V(r)$ was therefore assumed to have a short range which could be included in the definition of nuclear radius. If we are considering a process where Coulomb interaction is important, $V(r)$ will contain a positive term zZe^2/r, falling off even less rapidly than the centrifugal term. The Coulomb field can thus cause scattering at values of r so great that the centrifugal barrier does not need to be penetrated. Therefore, when Coulomb interaction is important, many partial waves must be considered, and the method becomes unsatisfactory.

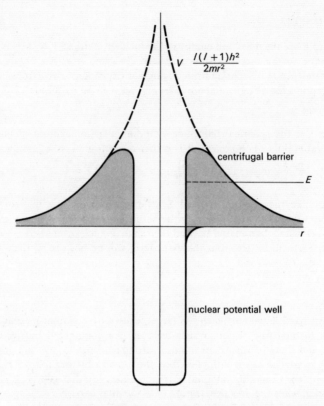

$V \quad \dfrac{l(l+1)\hbar^2}{2mr^2}$

centrifugal barrier

E

r

nuclear potential well

Figure 18 Centrifugal potential barrier illustrated diagrammatically as the result of superimposing a potential $l(l+1)\hbar^2/2mr^2$ on a nearly square potential well

5.4.2 Alternative treatment, using a classical model

The orbital angular momentum l of a partial wave is not easy to visualize in terms of the wave itself. We can, however, construct a simple model showing the significance of l in terms of the scattering process which gives rise to it. A particle of momentum p approaching a heavy target along a path with impact parameter r (see Figure 19) will have an angular momentum pr about an axis perpendicular to the plane of scattering. This is of course a crude model, but it serves to give physical significance to what is really a quantum-mechanical parameter.

The same model gives us an indication of how large the angular momentum can be. To give an angular momentum l, measured in units of $\hbar$, pr must be of order $l\hbar$. But for any nuclear interactions to occur, r must be less than the radius p of the target nucleus. A limit to l for nuclear interactions is therefore given by

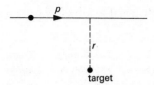

Figure 19 Classical model for angular momentum in scattering

$$l \leqslant \frac{p\rho}{\hbar} = \frac{\rho}{\lambda},$$

which is the same as equation **5.16**, except that we have used l instead of $\sqrt{\{l(l+1)\}}$, for the magnitude of the angular momentum.

The two alternative arguments thus show that our summations over l need be made only up to a limit set by the ratio

$$\frac{\text{Radius of target for nuclear interaction}}{\lambda \text{ of incident particles}}.$$

5.4.3 *Numerical values*

The preceding sections have shown that, in order to give the possibility of nuclear interaction in the partial wave with $l = 1$, λ must not exceed a limit set by the effective radius of the target nucleus. For scattering of neutrons, the effective radius is the potential radius of the nucleus itself, that is, the radius within which a neutron or proton begins to feel the effect of nuclear forces. This is slightly greater than the radius of the mass distribution, because of the non-zero range of nuclear forces. For scattering of larger particles – α-particles, for example – we should have to include an allowance for the radius of the incident particle.

For scattering of neutrons by nuclei of mass number A, we may use a potential radius

$$\rho = r_0 A^{\frac{1}{3}}, \qquad\qquad\qquad 5.25$$

where

$r_0 = 1\cdot30\,\text{fm}$

(1 fm $= 10^{-15}\,\text{m} = 10^{-13}\,\text{cm}$; see section 1.4). Using this value of ρ in equation **5.24**, with the value

$\hbar c = 197\cdot2\,\text{MeV fm}$,

we obtain Table 6 showing limits to c.m.s. momentum; also listed are the corresponding laboratory system momenta and energies, calculated relativistically in the one case $(n+p)$ that requires it.

Table 6 Momentum and Kinetic Energy Required to Give $l = 1$ in Neutron Scattering

Target	A	ρ/fm	c.m.s. momentum /(MeV/c)	Laboratory momentum /(MeV/c)	Kinetic energy (lab)/MeV
H	1	1·30	215	439	98
C	12	2·98	94	101	5·5
Fe	56	4·97	56	57	1·74
U	238	8·06	35	35	0·64

These figures point to an important conclusion: that there is no possibility of nuclear interaction in any partial wave except the S-wave, with $l = 0$, when the beam consists of neutrons of energy well below the limits set in the last column of the table. Thermal neutrons, with kinetic energy only a fraction of an electronvolt, satisfy this criterion with a very wide margin.

For proton beams, nuclear interactions will be predominantly in the S-wave, at energies up to a few million electronvolts for light elements; but Coulomb interaction may increase the relative importance of the P- and D-waves.

When interaction is in the S-wave only, the formulae for $f(\theta)$, $d\sigma/d\Omega$ and the cross-sections are particularly simple, since there is no longer any summation and the factor $2l + 1$ may be omitted.

5.5 Effects of spin

All the preceding discussion has assumed that the interacting particles have zero spin, or alternatively that their interactions are independent of spin. This is a simplification that allows us to treat many important problems and to make a start on others.

When there is spin, and the interaction depends on its direction, we can apply the simple theory to the different spin states, one at a time, using different values of the parameters to describe the different strengths of the interaction in the various states. To get a cross-section for a process in which spin orientations are not measured, we then have to sum over all the final spin states, and average over initial spin states.

When one of the interacting particles has spin, the S-wave interaction is unaffected, but if the incident energy is high enough to excite partial waves with $l \neq 0$, there may be spin–orbit interaction. It is then necessary to classify the partial waves according to the total angular momentum j, as well as according to l. For example, if a scattering process takes place through the formation of a compound nucleus with well-defined spin and parity, the important partial wave is the one with j equal to the required spin, and l even or odd according to whether the required parity is the same as the over-all intrinsic parity of the incident particles, or opposite to it.

The commonest cases of this type are scattering of neutrons, with spin one half, by target nuclei with spin zero, and in the field of elementary-particle physics, scattering of mesons with spin zero by protons with spin one half. In either case, the partial waves with $l \neq 0$ are divided into two according to whether j is $l+\frac{1}{2}$ or $l-\frac{1}{2}$. There is a further subdivision, according to whether or not the particle of spin $\frac{1}{2}$ has the direction of its spin changed. We use a new scattering amplitude $g(\theta)$ for the spin–flip scattering, keeping $f(\theta)$ for the non-flip scattering. The two processes obey slightly different rules; their contribution to the scattering add incoherently, so that the differential cross-section is given by

$$\frac{d\sigma}{d\Omega} = |f(\theta)|^2 + |g(\theta)|^2.$$ **5.26**

A full description of meson–proton scattering is given by Hughes (1971).

Last of all, we consider the possibility that both the interacting particles have non-zero spin. The simplest example of this is neutron–proton scattering, which is treated especially in Chapter 11. The methods of the present chapter are applied separately to the state in which neutron and proton have spins parallel, and to the state in which the spins are antiparallel. Different parameters are needed to describe the interactions in the two states, because nuclear forces are indeed spin-dependent. But the simple theory applies to each, with the appropriate parameters.

Spin-dependence of nuclear forces is treated, in its own right, in Chapter 12, along with polarization and some of the spin-dependent features of scattering which are not mentioned in the simple treatment.

Chapter 6
Resonance Scattering and Reactions

6.1 Compound nuclei and resonances

Reactions and scattering processes may take place through various mechanisms. One of them involves the formation of an intermediate state, or compound nucleus, which may have well-defined properties. When a process takes place predominantly through formation of a compound nucleus in a single state, the properties of this state determine many features of the process. The present chapter is concerned largely with the way in which the cross-sections of such processes are determined by the properties of the intermediate state.

To form a particular compound nucleus in a single excited level, the incident particle must have an appropriate energy, which may be calculated by the kinematic equations of Chapter 3, modified to give only a single product – the compound nucleus – with a Q-value determined by the excitation energy of the required level. We therefore expect the process to show the characteristics of a resonance, stimulation at the correct energy having more effect than stimulation at either higher or lower energy (see Figure 20). This sort of behaviour is typical of resonant systems of many types, though usually we think of frequency rather than energy as the independent variable. For example, an acoustic resonator or an electrical tuned circuit responds to stimulation at its characteristic frequency more than it does at either higher or lower frequency.

The resonant system, or compound nucleus, after it has been formed, may break up in one of various different ways. If it breaks up into two particles identical with those that interacted to form it, this is a special mechanism for scattering: elastic scattering if no energy is lost as γ-radiation or in excitation of either of the particles. But if the compound nucleus breaks up into different particles, a nuclear reaction has taken place. As a third possibility, the compound nucleus may retain its identity without breaking up, losing its energy of excitation by emitting a γ-ray: this would be called a capture reaction.

6.2 Resonance scattering, purely elastic

6.2.1 *The wave function of an unstable system*

The wave function of an unstable system may be written in the form

$$\Psi = \psi e^{-i\omega t} e^{-\frac{1}{2}t/\tau}, \qquad\qquad 6.1$$

where ψ is the space-dependent part of the wave function, normalized to make $|\psi|^2 = 1$ at $t = 0$, the moment at which the system is created. The term ω_0 is an angular frequency given by the energy E_0 of the system

$$\hbar\omega_0 = h\nu = E_0,$$

τ is the mean life, and the factor one half in **6.1** ensures that $|\Psi|^2$ decreases as $e^{-t/\tau}$.

According to the uncertainty principle, attempts to measure the energy of this short-lived system necessarily give results spread over a range Γ, which is called the level width. Γ is related to the mean life τ by the Heisenberg relation

$$\tau = \frac{\hbar}{\Gamma}.$$

If we substitute for τ and ω_0 in equation **6.1**, we get an expression for Ψ in terms of energies

$$\Psi = \psi e^{-i(t/\hbar)(E_0 - i\frac{1}{2}\Gamma)}.$$ **6.2**

We may therefore describe the essential properties of the unstable system by saying that it has a complex energy $E_0 - i\frac{1}{2}\Gamma$: the real part tells us the mean result of an ordinary measurement of the energy, while the complex part tells us the lifetime and the level width.

6.2.2 *Fourier integrals and Fourier components*

A periodic function of time may be expressed as a sum of Fourier components, each with frequency equal to a multiple of the fundamental frequency. A non-periodic function of time, however, needs an infinite number of components, with frequencies spread continuously over a wide range; it is therefore necessary to use a Fourier integral, instead of a simple Fourier sum. The standard method is to express a non-periodic function $f(t)$ as

$$f(t) = \int_{-\infty}^{\infty} a_\omega \, e^{-i\omega t} \, d\omega.$$

The amplitude a_ω per unit frequency interval of spectral components of frequency around ω is given by

$$a_\omega = \frac{1}{2\pi} \int_{-\infty}^{\infty} f(t) \, e^{i\omega t} \, dt.$$ **6.3**

Most students will be familiar with the application of this method to analysing non-periodic electrical or mechanical impulses, which are represented by real functions $f(t)$.

Now, however, we are going to apply the same method to the complex function which gives the time dependence of our wave function in equation **6.2**.

6.2.3 *Fourier components of the wave function of a resonance*

Let us put

$$f(t) = e^{-(it/\hbar)(E_0 - i\frac{1}{2}\Gamma)}.$$

The amplitude a_ω of the component with frequency ω is now given by equation **6.3** as

$$a_\omega = \frac{1}{2\pi} \int_0^\infty e^{-(it/\hbar)(E_0 - i\frac{1}{2}\Gamma)} e^{i\omega t}\, dt. \qquad\qquad \textbf{6.4}$$

The integral is taken from $t = 0$, instead of from $-\infty$, because we are regarding Ψ as zero during negative time, until at $t = 0$ the system is created and Ψ starts to be given by equation **6.2**.

Since the properties of the system are being described by the mean energy E_0 and the level width Γ, it is convenient to describe the component by its energy E instead of by its frequency ω. Substituting $E/\hbar$ for ω, equation **6.4** becomes

$$a_\omega = \frac{1}{2\pi} \int_0^\infty e^{(t/\hbar)\{i(E - E_0) - \frac{1}{2}\Gamma\}}\, dt$$

$$= \frac{1}{2\pi} \left[\frac{\hbar e^{(t/\hbar)\{i(E - E_0) - \frac{1}{2}\Gamma\}}}{i(E - E_0) - \frac{1}{2}\Gamma} \right]_0^\infty$$

$$= \frac{\hbar}{2\pi} \{i(E - E_0) - \tfrac{1}{2}\Gamma\}^{-1}. \qquad\qquad \textbf{6.5}$$

This means that our unstable system behaves like a spread of components, with energies spread around the central value E_0, each having amplitude proportional to

$$\frac{1}{i(E - E_0) - \frac{1}{2}\Gamma}$$

and therefore intensity proportional to

$$\left| \frac{1}{\{i(E - E_0) - \frac{1}{2}\Gamma\}} \right|^2 = \frac{1}{\{(E - E_0)^2 + \frac{1}{4}\Gamma^2\}}.$$

Thus, neglecting the spatial part of the wave function, we have expressed it as a quantum-mechanical superposition of energy eigenstates $e^{-i\omega t}$ with amplitude a_ω.

6.2.4 *The second significance of Fourier components*

Fourier methods closely analogous to the above may be used to describe the properties of a damped oscillatory electrical circuit: ω_0 represents the natural frequency, while τ represents the time constant for decay of an un-maintained oscillation. The reciprocal τ^{-1} gives the range of frequencies over which the natural oscillations appear to be spread. Students who are familiar with this problem will already know that the intensity of a particular Fourier component gives two things: (a) the amount of that component which must be included in the Fourier integral to represent the natural oscillations, and (b) the intensity of response of the circuit to external stimulation by a signal of frequency equal to that of the component.

Similarly in the nuclear case, the intensity of a Fourier component of energy E has two significances. It gives (a) the probability that a measurement of the energy of the system will yield the result E; that is, the spread of intensities gives the level width; and (b) the relative probability that the resonant system will be created by an incident particle which raises the energy to the value E.

6.2.5 *Energy dependence of the cross-section for formation of a resonance*

Having described the resonance in terms of a level width with significance (a), we shall now make use of significance (b). This tells us that the cross-section for formation of the resonance by an incident particle of given energy should be proportional to the intensity of the appropriate Fourier component. We may therefore write

$$\sigma(E) = \frac{\text{constant}}{\{(E - E_0)^2 + \tfrac{1}{4}\Gamma^2\}} \ . \qquad\qquad \textbf{6.6}$$

E and E_0 were originally introduced as energies measured from an unspecified zero in the rest frame of the compound nucleus, which is the same as the c.m.s.: only their difference appears in the right-hand side of equation **6.6**, so we are at liberty to choose any zero we like for the applications of this equation. For convenience, we shall let E be the kinetic energy of the incident particle, so that E_0 is the incident kinetic energy required to excite the resonance at its peak. If we put $E = E_0$ in equation **6.6**, we obtain for the peak value of the cross-section

$$\sigma(E_0) = \text{constant} \times \frac{4}{\Gamma^2}. \qquad\qquad \textbf{6.7}$$

6.2.6 *The cross-section put on an absolute scale*

Now let us suppose that this resonance, having been formed, can do nothing but break up elastically into the particles from which it was formed. Its formation and break-up is then behaving as a special mechanism for elastic scattering. This special type of elastic scattering must obey the wave-mechanical rules of Chapter 5, which were in no way conditional on the mechanism. In particular, the maximum possible value of the elastic scattering cross-section is given for the lth partial wave by

$$\sigma_{\text{elastic(max)}} = 4\pi \lambdabar^2 (2l + 1). \qquad\qquad \textbf{[5.16]}$$

If the scattering is perfectly elastic, it is reasonable to assume that this maximum possible value will just be reached at the peak of the resonance. We may then combine equations **5.16** and **6.7** to find that the constant of equation **6.6** and **6.7** is

$$\pi \lambdabar^2 (2l+1)\Gamma^2.$$

Equation **6.6** may therefore be rewritten as

$$\sigma(E) = \pi\lambdabar^2(2l+1)\frac{\Gamma^2}{(E-E_0)^2 + \frac{1}{4}\Gamma^2}. \qquad \textbf{6.8}$$

This is the simplest of the Breit–Wigner formulae, which describe the energy dependence of processes going through a single resonance. They all contain

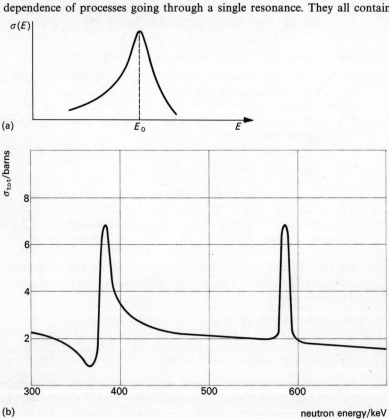

Figure 20 Resonance in cross-section for purely elastic scattering. (a) Ideal Breit–Wigner resonance; (b) observed total cross-section for scattering of neutrons by sulphur

the resonance term $\{(E-E_0)^2 + \frac{1}{4}\Gamma^2\}^{-1}$; the other terms depend on the process described (see sections 6.3.4 and 6.3.5). Those given in equation **6.8** apply only to the case of pure elastic scattering. An example showing the characteristic shape of the resonance term is shown in Figure 20(a). For comparison with this idealized case, Figure 20(b) shows of the peaks in the total cross-section for scattering of neutrons by sulphur, which were observed experimentally by Peterson, Barschall and Bockelman (1950). The peak at 585 keV may be described as having a nearly ideal resonance shape superposed on a background which varies slowly with energy. The peak at 375 keV, however, has a minimum on the low-energy side and a shoulder on the high-energy side: these are attributed to interference between resonance and potential scattering (see section 7.2.2).

Here, and in what follows as far as section 6.3.7, we are ignoring effects of spin, and assuming that only one value of l can lead to formation of a given resonance.

6.3 Resonance scattering, with competing reactions

6.3.1 Channels

Often a compound nucleus or resonant system can break down in several different ways. Correspondingly, it may be possible to form a given resonant system from several different sets of initial particles. These different sets of initial or final particles are called channels, entrance channels for the initial particles and exit channels for the final particles. In the case of elastic scattering, the exit channel is the same as the entrance channel. For a reaction the exit channel is different.

As an example, we may quote the case of carbon-12, which has many excited states. One of these, at 19·3 MeV, has been observed to be formed through the following two entrance channels:

$^{11}_{5}B + ^{1}_{1}H$, with proton having kinetic energy 3·6 MeV,

$^{12}_{6}C + \gamma$, with γ-ray energy 19·3 MeV.

Once formed, through either of these entrance channels, it can break up by any of the seven exit channels shown in Figure 21.

6.3.2 Cross-sections and partial widths of exit channels

When one entrance channel can lead to more than one exit channel, each process will have its own cross-section; the cross-section for a particular exit channel will be a definite fraction of the total cross-section, the fraction being given by the ratio

$$\frac{\text{Probability per second that the compound nucleus breaks up into this channel}}{\text{Total probability per second that it breaks up somehow}}.$$

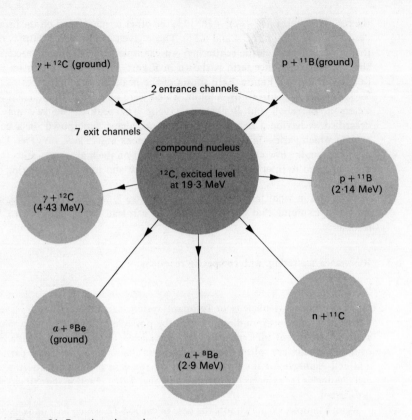

Figure 21 Reaction channels

The total probability per second for break-up of the compound nucleus is $\Gamma/\hbar$, where Γ is the level width. If we introduce partial widths $\Gamma_1, \Gamma_2, ..., \Gamma_n$, for exit channels 1, 2, ..., n, such that the probability per second of break-up of the compound nucleus into channel i is $\Gamma_i/\hbar$, the sum of these probabilities must be $\Gamma/\hbar$. Thus the sum of the partial widths is the level width Γ

$$\sum_{i=1}^{n} \Gamma_i = \Gamma.$$

The cross-sections are then in the ratio of the widths, according to

$$\sigma_1 : \sigma_2 : ... : \sigma_n : \sigma_{\text{total}} = \Gamma_1 : \Gamma_2 : ... : \Gamma_n : \Gamma.$$

Γ is called the level width because it tells us the width of the peak in a plot of cross-section or Fourier intensity against energy. The partial widths $\Gamma_1, ..., \Gamma_n$ are not widths in this sense. They are merely parts of the level width, obtained by dividing it up in proportion to the cross-sections.

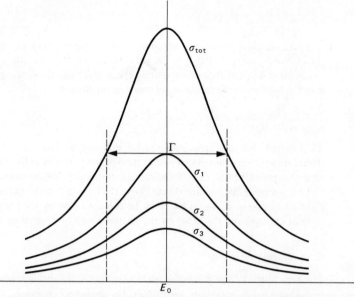

Figure 22 Total and partial cross-sections for three exit channels formed through a single resonance of width Γ. The partial cross-sections have the same width, but magnitudes in the ratio of the partial widths: $\sigma_1 : \sigma_2 : \sigma_3 : \sigma_{tot} = \Gamma_1 : \Gamma_2 : \Gamma_3 : \Gamma$

We should therefore expect that when several exit channels are available, it will be necessary to modify equation **6.8**, which gave the cross-section for elastic scattering through a resonance. The energy dependence of the cross-section comes in the formation of the resonance, not in its subsequent break-up; therefore the term in the denominator remains unchanged, with Γ as the level width, the factor that determines the sharpness of the resonance. The cross-sections for the different exit channels (of which elastic scattering is one) have to be in the ratio of their partial widths; therefore we expect the cross-section for each exit channel to contain the partial width instead of one of the factors Γ in the numerator. Figure 22 illustrates the energy dependence of the cross-sections for three exit channels formed through a single resonance.

6.3.3 *Partial width of entrance channel*

The above argument, however, does not go far enough: we cannot label channels as 'entrance' or 'exit' channels in any absolute sense. At a constant total energy in the c.m.s., a family of nuclear reactions exhibits reversibility. The entrance channels of the direct reactions are the exit channels of the reverse reactions, and vice versa. The cross-section of a reaction is related to that of the reverse reaction by the reciprocity theorem, which is given by Blatt and Weisskopf (1952, p. 336) in the form

$$\sigma_{b \to a} = \frac{\lambda_b^2}{\lambda_a^2} \sigma_{a \to b}. \qquad \textbf{6.9}$$

The term λ_b is the value of λ which gives the required c.m.s. energy when b is the entrance channel, and λ_a is the value when a is the entrance channel.

Equation **6.9**, and the corresponding relation for reactions between channels c and a, lead us to the following expression for the ratio

$$\frac{\sigma_{b \to a}}{\sigma_{c \to a}} = \frac{\lambda_b^2}{\lambda_c^2} \cdot \frac{\sigma_{a \to b}}{\sigma_{a \to c}} = \frac{\lambda_b^2}{\lambda_c^2} \frac{\Gamma_b}{\Gamma_c}. \qquad \textbf{6.10}$$

This means that the cross-sections for creating a given exit channel from different entrance channels are in the ratio of the partial widths of the entrance channels, provided that each is measured in terms of λ^2 for the incident particles.

Thus we reach the conclusion that both the factors Γ in the numerator of the right-hand side of equation **6.8** must be replaced, one by the partial width of the exit channel and the other by the partial width of the entrance channel.

6.3.4 *Breit–Wigner Formulae*

The conclusion of the preceding section is that if channel a represents a pair of particles a + A, and so on for b and c, the cross-section for the reaction

a + A $\longrightarrow$ c + C

at energy E will be

$$\sigma_{a \to c}(E) = \pi \lambda^2 (2l+1) \frac{\Gamma_a \Gamma_c}{(E - E_0)^2 + \frac{1}{4}\Gamma^2}. \qquad \textbf{6.11}$$

When the two channels are the same, the process is elastic scattering, e.g.

a + A $\longrightarrow$ a + A,

with cross-section

$$\sigma_{a,\text{ elastic}}(E) = \pi \lambda^2 (2l+1) \frac{\Gamma_a^2}{(E - E_0)^2 + \frac{1}{4}\Gamma^2}. \qquad \textbf{6.12}$$

If we group together all reactions (i.e. all processes except elastic scattering), we obtain what was called in Chapter 5 the inelastic cross-section. This is

$$\sigma_{a,\text{ inelastic}}(E) = \pi \lambda^2 (2l+1) \frac{\Gamma_a(\Gamma - \Gamma_a)}{(E - E_0)^2 + \frac{1}{4}\Gamma^2}. \qquad \textbf{6.13}$$

But if we take all processes, including elastic scattering, we have the total cross-section for channel a at energy E

$$\sigma_{a,\text{ total}}(E) = \pi \lambda^2 (2l+1) \frac{\Gamma_a \Gamma}{(E - E_0)^2 + \frac{1}{4}\Gamma^2}. \qquad \textbf{6.14}$$

The four equations **6.11-14** are further examples of Breit–Wigner formulae, applicable to competing processes with fixed values of partial width.

All the processes going through a given resonance have cross-sections containing the same resonant term $\{(E-E_0)^2 + \frac{1}{4}\Gamma^2\}^{-1}$. They have equivalent-looking terms $\lambda^2\,(2l+1)$ which fix different over-all scales for the cross-sections in the different entrance channels. The remaining factors are as given in Table 7.

Table 7 Relative Magnitudes of Cross-Sections for Reactions between Three Channels (All in Units of $\lambda^2(2l+1)$ for Entrance Channel)

| | Entrance channel | | |
	a	b	c
Exit channel			
a	Γ_a^2	$\Gamma_b\,\Gamma_a$	$\Gamma_c\,\Gamma_a$
b	$\Gamma_a\,\Gamma_b$	Γ_b^2	$\Gamma_c\,\Gamma_b$
c	$\Gamma_a\,\Gamma_c$	$\Gamma_b\,\Gamma_c$	Γ_c^2
total	$\Gamma_a\,\Gamma$	$\Gamma_b\,\Gamma$	$\Gamma_c\,\Gamma$

6.3.5   *Maximum values*

The above results are all consistent with those obtained in Chapter 5 for the maximum values of cross-sections. At resonance, equation **6.12** gives the peak value of the elastic scattering cross-section for channel a as

$$\sigma_{a,\,\text{elastic}}(\text{res}) = \pi\lambda^2\,(2l+1)\frac{4\Gamma_a^2}{\Gamma^2}. \qquad \textbf{6.15}$$

In the case of pure elastic scattering, with all other channels closed, $\Gamma_a = \Gamma$, and this reaches its permitted maximum of $4\pi\lambda^2\,(2l+1)$.

The total cross-section at resonance is given by equation **6.14** as

$$\sigma_{a,\,\text{total}}(\text{res}) = \pi\lambda^2\,(2l+1)\,\frac{4\Gamma_a\,\Gamma}{\Gamma^2}, \qquad \textbf{6.16}$$

which reaches the same maximum value under the same conditions.

Finally, equation **6.13** gives, for the inelastic cross-section at resonance,

$$\sigma_{a,\,\text{inelastic}}(\text{res}) = \pi\lambda^2\,(2l+1)\frac{4\Gamma_a(\Gamma-\Gamma_a)}{\Gamma^2}. \qquad \textbf{6.17}$$

This has its greatest possible value when

$$\Gamma_a = \tfrac{1}{2}\Gamma,$$

making $\sigma_{\text{inelastic}}(\text{max}) = \pi\lambda^2\,(2l+1),$

in agreement with the maximum value deduced in Chapter 5.

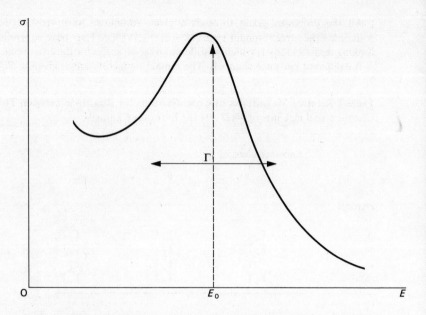

Figure 23 Skew resonance peak obtained with E_0 not much greater than Γ ($E_0 = 1 \cdot 5\,\Gamma$ in this example)

6.3.6 Energy dependence

We have drawn attention to the resonant term $(E - E_0)^2 + \frac{1}{4}\Gamma^2$ as determining the shape of a plot of cross-section against energy, but we must now remember that it is not the only energy-dependent term: λ^2 varies as E^{-1} in non-relativistic conditions. This is a much slower and smoother variation than that of the resonant term; but if the level width Γ is not much smaller than the resonance energy E_0, λ^2 will have significantly different values on the two sides of the resonance peak. We may thus expect to find a skew (see Figure 23) in the plot of cross-section against energy, and we must allow for the skewness when fitting measurements to a broad resonance.

The above applies to cases in which the partial widths are constants, independent of energy. Although we have assumed them to be constant up to now, we shall in section 6.5 meet the case of neutron interactions in which the energy dependence of partial widths serves to cancel, in part or in whole, the energy dependence of λ^2.

6.3.7 Effects of spin

So far we have ignored the effects of possible spin of the particles participating in a resonant process. Our results would describe correctly the interactions of particles with spin, provided that a channel was defined as a set of particles with specified orientation of spins, and care was taken to choose channels between which angular momentum was conserved.

A more useful approach, however, is to define a channel in terms of the particles making it up, and allow for all possible orientations of their spins. The general case, where all the particles have non-zero spin, and l can be non-zero, can become extremely complicated. As a first simplification, we shall assume that elastic scattering with change of orbital angular momentum is negligible. (Conservation of parity would require l to change by two units, which could be made up from spin angular momentum, but this is very unlikely at moderate energies.) Also we shall assume that our resonant process is going through a compound nucleus with definite angular momentum J.

Let the incident particle have spin $\mathbf{s}$, and the target particle spin $\mathbf{I}$; these may be combined to make a channel spin

$$\mathbf{S} = \mathbf{s} + \mathbf{I}.$$

The total angular momentum $\mathbf{J}$ must be the vector sum of $\mathbf{S}$ and the orbital angular momentum l;

$$\mathbf{J} = \mathbf{S} + l$$

Our cross-sections, with spin ignored, have all contained a factor $2l+1$ for the number of possible orientations of l, with z-component $l, l-1, \ldots, -l$.

If the initial particles have spins s and I, the probability of their being in a state of given channel spin S is

$$g(S) = \frac{(2S+1)}{(2s+1)(2I+1)}. \qquad \qquad \textbf{6.18}$$

Further, the probability that a channel spin S, with a particular orbital angular momentum l, will make up the required total angular momentum J is

$$g(J) = \frac{(2J+1)}{(2l+1)(2S+1)}.$$

Therefore, the over-all probability of making the required J, with a given l, is

$$g(S)\,g(J) = \frac{(2J+1)}{(2l+1)(2s+1)(2I+1)}. \qquad \qquad \textbf{6.19}$$

To allow for the spins in our formulae for the cross-sections, we must include the factor $g(S)\,g(J)$. This has the effect of replacing the factor $(2l+1)$ by

$$\frac{(2J+1)}{(2s+1)(2I+1)}.$$

The term in l has disappeared from the multiplicity factor, but we must remember that the process cannot go unless there exists a value of l which can give to the compound nucleus the necessary total angular momentum J, and give it the correct parity if this is defined.

We conclude that equation **6.11** for the cross-section for a resonant process between channels a and c, must be replaced in the case of non-zero spins by

$$\sigma_{a \to c}(E) = \pi \lambda^2 \frac{(2J+1)}{(2s+1)(2I+1)} \cdot \frac{\Gamma_a \Gamma_c}{(E-E_0)^2 + \frac{1}{4}\Gamma^2}.$$

Corresponding adjustments may be made to the various other cross-sections.

In the special case of interactions of slow neutrons, we are limited to $l = 0$. Since $s = \frac{1}{2}$, J can only have two values, namely $I + \frac{1}{2}$ and $I - \frac{1}{2}$.

Other special cases lead to their own particular simplifications. For example, the scattering of fast neutrons by α-particles (with $I = 0$) has

$$\frac{(2J+1)}{(2s+1)(2I+1)} = J + \frac{1}{2}.$$

6.4 Wave-mechanical description of resonant processes

6.4.1 *The transition amplitude obtained from the cross-sections*

The conclusions of the preceding sections may be translated into the language of Chapter 5 by fitting equations **6.12** and **6.13** to equations **5.10** and **5.11** respectively. If we consider scattering and reactions taking place through a single resonant level, formed by one partial wave, we may drop the suffix l from T_l, η_l and δ_l, writing T, η and δ. The transition amplitude $T = T_l$ is determined by the parameters of the level; $|T|^2$ is given by the equations for σ_{elastic} (equations **6.12** and **5.10**), which fit if

$$|T|^2 = \frac{\frac{1}{4}\Gamma_a^2}{(E-E_0)^2 + \frac{1}{4}\Gamma^2}. \qquad\qquad 6.20$$

A second number, required to fix the real and imaginary parts of the complex T, is provided by the elasticity η. To make the equations for $\sigma_{\text{inelastic}}$ (equations **6.13** and **5.11**) fit, we must have

$$1 - \eta^2 = \frac{\Gamma_a(\Gamma - \Gamma_a)}{(E-E_0)^2 + \frac{1}{4}\Gamma^2}. \qquad\qquad 6.21$$

6.4.2 *The resonant transition amplitude plotted in the complex plane*

Ideal case. Equations **6.20** and **6.21** fix the point representing T in the complex plane, as plotted in a diagram of the type shown in Figure 16 (p. 69). The representative point is fixed by being at a distance $|T|$ from the origin, and at a distance $\frac{1}{2}\eta$ from the point $\frac{1}{2}i$, as shown in Figure 24. Where this gives two alternatives, we choose the one that gives the real part of T_l the same sign as $E_0 - E$. A little algebra shows that the angle ϕ of Figure 24 is given by

$$\tan \phi = \frac{E_0 - E}{\frac{1}{2}\Gamma}.$$

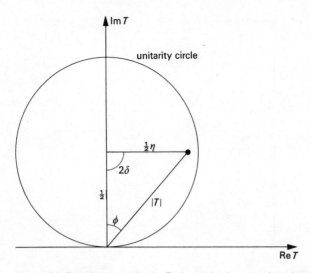

Figure 24 Transition amplitude T plotted in the complex plane, near a resonance $\tan \phi = (E_0 - E)/\frac{1}{2}\Gamma$

and that T may be written as a complex quantity,

$$T = \frac{\frac{1}{2}\Gamma_a\{(E_0 - E) + i\frac{1}{2}\Gamma\}}{(E - E_0)^2 + \frac{1}{4}\Gamma^2}$$

$$= \frac{\frac{1}{2}\Gamma_a}{(E_0 - E) - i\frac{1}{2}\Gamma}. \tag{6.22}$$

As the energy increases, the representative point moves anticlockwise round a circle of radius $\Gamma_a/2\Gamma$, with its centre on the imaginary axis at the point $i\Gamma_a/2\Gamma$. Figure 25 shows that, as the point goes round the circle, both the phase shift δ and the elasticity η change. At the resonance, T is purely imaginary, η is a minimum, and both elastic and inelastic cross-sections go through peak values. The phase shift δ goes through 90°, unless Γ_a/Γ is less than one half, in which case δ goes through 0° (see Figure 25b). In fact, for highly inelastic resonances, with $\Gamma_a \ll \frac{1}{2}\Gamma$ the phase shift δ is not a particularly useful parameter.

Real cases. Real cases are not so simple as has been suggested above; the simple theory given has been concerned with transitions taking place entirely through the formation of a single resonant level in a compound nucleus. There are real cases which approximate quite closely to this, but in practice it is usually necessary to allow for the presence of further levels at more or less remote energies, or to allow for some non-resonant scattering.

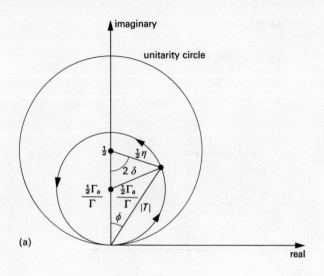

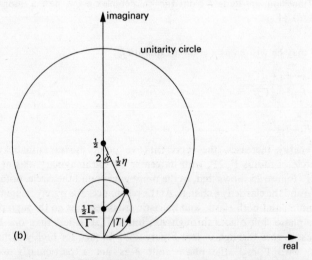

Figure 25 Idealized diagrams of scattering amplitude T as a function of energy for a single resonance: (a) with $1 > \Gamma_a/\Gamma > \frac{1}{2}$; (b) with $\Gamma_a/\Gamma < \frac{1}{2}$

A succession of levels causes the plot of T to describe a set of circles as shown in Figure 26(a). Non-resonant scattering, through the potential of the target nucleus affecting the incident particle without formation of a compound nucleus, gives a contribution to the transition amplitude which varies relatively slowly with incident energy. The result is to modify the ideal single-level plot of Figure 25(b), giving one of the less symmetrical diagrams of Figure 26.

92   Resonance Scattering and Reactions

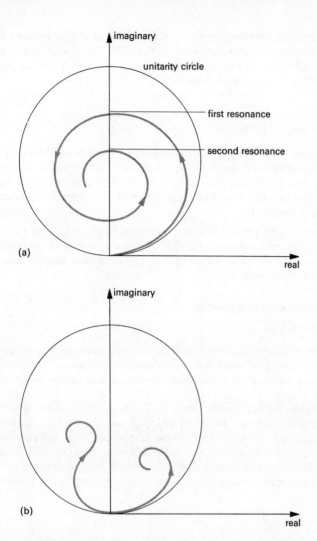

Figure 26 Scattering amplitudes plotted in the complex plane (a) for two resonances, (b) for one resonance in the presence of two alternative forms of potential scattering

6.4.3 *The Fourier method re-examined*

The complex expression **6.22** for the transition amplitude T near a resonance was obtained from the expressions for the elastic and inelastic cross-sections, both of which contain the Breit–Wigner denominator $(E-E_0)^2+\frac{1}{4}\Gamma^2$; the latter first appeared as the square of the modulus of the amplitude a_ω of the

Fourier component with frequency $\omega = E/\hbar$. The fact that a_ω is complex has had no influence on the argument since section 6.2.3. But if we now go back to equation **6.5** for a_ω as a complex quantity, and compare it with equation **6.22**, we see that T and a_ω are related by

$$T = \frac{\pi \Gamma_a}{i\hbar} a_\omega.$$

From this relation we deduce two important things:

(a) The argument of section 6.2 is still valid when extended to cover the general case of elastic scattering in the presence of reactions. Our device for extending it by using appropriate partial widths for the different channels could have been invoked at the start, if we had not preferred to work towards the general case by way of the simpler.

(b) Apart from the 90° phase difference represented by the factor i, the Fourier component gives the transition amplitude in phase as well as in magnitude. This fact, once proven, offers an alternative, and sometimes simpler, way of calculating the cross-sections.

6.5 Resonance reactions of slow neutrons

6.5.1 *Reactions of slow neutrons*

We have been discussing the general theory of reactions and scattering processes going through single resonant levels of a compound nucleus. The general approach was adopted in order that the conclusions might have wide application.

One important field of application of the Breit–Wigner formulae is in the reactions of slow neutrons. Nuclei of any type can absorb slow neutrons, thereby being transformed into nuclei of the next isotope of the same element. These are created in short-lived excited states which can decay by exit channels of the following types: (a) the elastic channel, which gives re-emission of the neutron as a scattered neutron; (b) the reaction channels, leading to processes like

$$^{10}_{5}\text{B} + ^{1}_{0}\text{n} \longrightarrow {}^{4}_{2}\text{He} + {}^{7}_{3}\text{Li};$$

(c) the capture channel, which is a short-hand name for the process in which the compound nucleus remains intact, usually losing its energy of excitation by emitting a γ-ray.

Some of the reactions will be considered, for their own interest, in Chapter 8. Now, however, we shall obtain the particular versions of the Breit–Wigner equation which are appropriate for describing their resonance-like behaviour.

6.5.2 *Partial width of neutron channel*

Density-of-states factor. A system consisting of a neutron emerging from inter-action with a nucleus does not have a true continuum of possible states; the momentum of the neutron has a limited number of possible values, each giving one state of the system. We are not interested in the absolute number of these states, nor even in the absolute number per unit energy; but we do need to know how the number per unit energy depends on the velocity of the neutron.

The number of states per unit energy appears as the density-of-states factor dN/dE in the general formula for the probability per second of a transition. This formula, sometimes called Fermi's golden rule number 2, is quoted by Fermi (1950, p. 143) in the form

$$w = \frac{2\pi}{\hbar} |H|^2 \frac{dN}{dE}.$$

6.23

(see also Schiff, 1968, p. 199). Here w is the probability per second that a transition will be caused by the perturbation to which the matrix element H refers.

To calculate how the density-of-states factor depends on velocity, in the case of a neutron interacting with a nucleus, we imagine that the neutron is confined in a box with z-dimensions L, and that it is moving along the z-axis with velocity v. The possible states are those for which the de Broglie waves of the neutron form a standing-wave system in the box; that is, the momentum p of the neutron is limited to those values which give

$$\frac{h}{p} = \lambda = \frac{L}{n},$$

where n is an integer. In the non-relativistic conditions applicable to slow neutrons, we may put

$$v = \frac{p}{m} = \frac{h}{m\lambda} = n \frac{h}{mL}.$$

The kinetic energy E is

$$E = \frac{mv^2}{2} = \frac{n^2 h^2}{2mL^2}.$$

Each value of n represents one state, so we may write

$$\frac{dN}{dE} = \frac{dn}{dE} = \frac{nh^2}{mL^2} = \frac{h}{L} v.$$

Velocity dependence of partial width. We are concerned, not with a neutron confined in a box, but with a neutron having its momentum determined by interactions inside the compound nucleus. The two situations have so much in common that we may take the density-of-states factor for the real situation as being proportional to v; we do not need to bother about whether to take literally the factor h/L.

Thus the probability of the compound nucleus decaying by elastic re-emission of a neutron contains an extra factor proportional to the velocity of the emitted neutron. This is the same as the velocity of the incident neutron, so we can accommodate the extra factor by saying that the partial width for the neutron channel, Γ_n, is proportional to λ^{-1}.

6.5.3 Breit–Wigner equation for elastic scattering of neutrons

The above argument applies to the incoming neutrons as well as to the outgoing ones, so the cross-section for elastic scattering contains two velocity-dependent factors Γ_n. These combine with the explicit factor λ^2 in equation **6.10**† to give

$$\sigma_{\text{elastic}} = \frac{\pi \lambda^2 \, \Gamma_n^2}{(E - E_0)^2 + \frac{1}{4}\Gamma^2}$$

$$= \frac{\pi \lambda_{\text{res}}^2 \Gamma_{n,\,\text{res}}^2}{(E - E_0)^2 + \frac{1}{4}\Gamma^2}. \qquad \textbf{6.24}$$

Here we have replaced the constant product $\lambda \Gamma_n$ by the product of the fixed values, λ_{res} and $\Gamma_{n,\,\text{res}}$, for λ and Γ_n at the resonant energy.

Alternatively, a reduced width may be used. This is the value of Γ_n taken at a fixed neutron energy, usually one electronvolt. When a reduced width is put in the formula, the value of λ for this fixed energy must be inserted.

6.5.4 Breit–Wigner equation for reactions of slow neutrons

For reactions, as distinct from scattering, we must use a fixed partial width Γ_r for the exit channel, and a velocity-dependent Γ_n for the entrance channel. This gives, for the reaction cross-section,

$$\sigma_{\text{reaction}} = \frac{\pi \lambda^2 \Gamma_n \, \Gamma_r}{(E - E_0)^2 + \frac{1}{4}\Gamma^2}$$

$$= \frac{\pi \lambda \lambda_{\text{res}} \, \Gamma_{n,\,\text{res}} \, \Gamma_r}{(E - E_0)^2 + \frac{1}{4}\Gamma^2}. \qquad \textbf{6.25}$$

6.6 Summary of Breit–Wigner equations

Equation **6.25** is a Breit–Wigner equation with energy dependence lying between that of the other two principal Breit–Wigner equations. We now list all three together for comparison, with constant terms first:

---

†We omit the factor $2l + 1$ since we are dealing with slow neutrons, which can interact only in the S-wave.

(a) Basic equation for any process $a \longrightarrow c$ with constant partial widths (useful as a step in the derivation of others, and as an approximation in the case of a narrow resonance)

$$\sigma_{a \to c} = \pi \Gamma_a \Gamma_c \frac{\lambda^2}{(E - E_0)^2 + \frac{1}{4}\Gamma^2} \quad \text{(from 6.11)}.$$

(b) Reaction cross-section for slow neutrons

$$\sigma_{\text{reaction}} = \pi \lambda_{\text{res}} \Gamma_{\text{n, res}} \Gamma_{\text{r}} \frac{\lambda}{(E - E_0)^2 + \frac{1}{4}\Gamma^2} \quad \text{(from 6.25)}.$$

(c) Elastic scattering cross-section for slow neutrons

$$\sigma_{\text{elastic}} = \pi \lambda_{\text{res}}^2 \Gamma_{\text{n, res}}^2 \frac{1}{(E - E_0)^2 + \frac{1}{4}\Gamma^2} \quad \text{(from 6.24)}.$$

Chapter 7
Reaction Mechanisms

7.1 Review

7.1.1 *Models*

It is now time to consider in rather more detail the actual mechanisms of nuclear processes. One particular mechanism, namely compound-nucleus formation, has been used in Chapter 6 as the basis for discussion of resonant processes. This mechanism now needs discussion in its own right, as do the direct processes of stripping and pick-up, and the intermediate processes described by the optical model.

In terms of a simple model, with the target nucleus consisting of a number of independent nucleons bound together by nuclear forces, a reaction may start

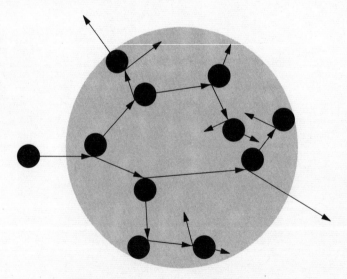

Figure 27 Nucleon–nucleus collision considered as a succession of nucleon–nucleon collisions. The incident energy is eventually spread over the nucleus as thermal energy, apart from that which is carried off by escaping particles

with an incident nucleon colliding with one of the target nucleons. Some of the energy of this collision will be transferred immediately to the rest of the target nucleus, through the nuclear forces binding the struck nucleon to it. The rest of the kinetic energy will be carried on through the nucleus by the incident and the struck nucleons, until they themselves have second-generation collisions with other target nucleons (see Figure 27). Third and later generations of collisions then serve to distribute the available energy over the whole nucleus. At any stage in the process, particles may escape from the nucleus. In the early stages, they may do so with relatively high kinetic energy, but once the energy has been uniformly distributed throughout the nucleus, concentration of a large energy on one nucleon or group of nucleons is unlikely, and the disintegration occurs by a process more like evaporation.

Clearly a full description of a nuclear reaction, stage by stage, would be very difficult, even in terms of this simple model. Therefore we usually give a partial description in which the nucleons of the target are treated, not as individuals, but as a homogeneous lump of nuclear matter.

This lump may scatter an incident particle through forces described by an ordinary spherically symmetric potential. Alternatively it may capture the incident particle, robbing it of the energy which identified it as part of the system of incidental elastically scattered waves. After an interval of time long compared with the nuclear radius$/c$, this energy will have been shared among all the nucleons present, and the resultant compound nucleus will have forgotten how it was formed. It will then break up by one of the exit channels characteristic of its own particular properties.

In contrast to the above mechanism, which is compound nucleus formation with or without early escape of an incident or a recoil nucleon, there are direct processes, in which the final state is formed directly from the initial state without going through any intermediate state. An extreme example of this is potential scattering (in contrast to compound elastic scattering where the original particles are re-formed after spending a period as a compound nucleus). When the incident particle is a deuteron, one of its constituent nucleons may go on relatively undisturbed, while the other is detached and enters the target. This so-called stripping reaction, and its converse in which an incident nucleon picks up a second nucleon to form a deuteron, are the best-known direct reactions. They form the subject of section 7.6.

7.1.2   *Range of validity of different models*

The various models of nuclear processes should be seen, not as competing for places in single list of over-all merit, but rather as alternative descriptions each with its own range of validity and usefulness.

When the compound nucleus is formed in a state with well-defined properties, we have resonance-like behaviour, with cross-sections governed by Breit–Wigner formulae. At low energies resonances tend to be well separated, so that the cross-section at a given energy is given by the sum of a small number of Breit–Wigner terms. But at medium energies resonances tend to overlap and there are many exit channels; then the sum of the Breit–Wigner terms is no longer adequate, and the optical model provides a better description of the cross-sections, averaging them over a band of energies wide enough not to be dominated by any single resonance. Thus the resonance model tends to apply at low energies, its validity decreasing as the energy increases and the optical model becomes better.

At still higher energies, we reach the range of validity of the models involving direct processes and interactions with single nucleons, the rest of the nucleus becoming less and less important.

7.2   Resonance and potential scattering

7.2.1   *The logarithmic derivative f*

In order to proceed to the evaluation of cross-sections in terms of assumptions about processes occurring inside a target nucleus, it is necessary to develop a little further the general theory of scattering as presented in Chapter 5.

We introduce a radial wave function $u(r)$, equal to $rR(r)$ in the notation of equation **5.18**. There is a radial wave function $u_l(r)$ for each partial wave, but for simplicity we shall drop the suffix l and consider only the partial wave with $l = 0$. If $l = 0$, equation **5.2** gives the radial wave function in the region where the potential $V(r) = 0$, as

$$u(r) = \frac{1}{2ik} \left(\eta e^{2i\delta} e^{ikr} - e^{-ikr} \right).$$

This specifies the radial dependence of the total wave function, which is the wave function of the incident plane waves, modified by inclusion of the factor $\eta e^{2i\delta}$ to describe the effect of the target nucleus on the outgoing spherical partial waves. It should hold for all values of r not less than the radius a of the target nucleus.

Another useful parameter is the logarithmic derivative of $u(r)$, which may be evaluated at the surface of the target nucleus as

$$f = \left[\frac{r}{u} \frac{du}{dr} \right]_{r\,=\,a}$$

$$= \frac{ika(\eta e^{2i\delta} e^{ika} + e^{-ika})}{(\eta e^{2i\delta} e^{ika} - e^{-ika})}.$$

This may be rewritten to give $\eta e^{2i\delta}$ in terms of the logarithmic derivative f,

$$\eta e^{2i\delta} = \frac{f + ika}{f - ika} e^{-2ika}. \tag{7.1}$$

The cross-sections for elastic scattering and for inelastic processes are determined by η and δ. Equation **7.1** therefore shows that the cross-sections for a particular incident momentum are determined by the logarithmic derivative f of the radial wave function at the surface of the target nucleus. To calculate the cross-sections we do not need to know everything about the processes occurring inside the nucleus, only their effect on f.

7.2.2 Cross-section for elastic scattering

Equation **5.9** gives the elastic cross-section in terms of a transition amplitude T_l, which is defined for each partial wave as $(\eta_l e^{2i\delta_l} - 1)/2i$. We may rewrite equation **5.9** for the $l = 0$ partial wave as follows, substituting from equation **7.1**,

$$\sigma_{\text{elastic}} = \pi \lambda^2 \left| 2iT \right|^2$$

$$= \pi \lambda^2 \left| \eta e^{2i\delta} - 1 \right|^2$$

$$= \pi \lambda^2 \left| A_{\text{res}} + A_{\text{pot}} \right|^2,$$

where $\quad A_{\text{res}} = \dfrac{-2ika}{f - ika}$ $\qquad\qquad$ **7.2**

and $\quad A_{\text{pot}} = e^{2ika} - 1.$ $\qquad\qquad$ **7.3**

Apart from a factor $\frac{1}{2}ie^{2ika}$, A_{res} and A_{pot} defined in this way together make up T, the transition amplitude. A_{res} is the part of the amplitude that depends on the shape of the wave function at the nuclear surface; the suffix 'res' has become conventional to indicate the amplitude which depends on the properties of the nucleus, whether or not these lead to resonance-like behaviour in a particular case. If the nucleus behaved like a hard sphere, the wave function would be zero inside and f would be infinite; then the only contribution to the transition amplitude would be given by A_{pot}, which is known as the amplitude for potential scattering or hard-sphere scattering. A_{pot} depends on the incident momentum and on the radius of the target nucleus, but not on the details of what happens inside it.

7.2.3 Cross-section for reaction

If there are inelastic processes occurring; for example, reactions taking place through formation of a compound nucleus, the combined cross-section for them must be related to the elasticity η by equation **5.10**, which for $l = 0$ gives

$$\sigma_{\text{inelastic}} = \pi \lambda^2 (1 - \eta^2).$$

By equations **7.1** and **7.2**, η is given as

$$\eta = \left| \frac{f + ika}{f - ika} \right| = \left| 1 - A_{\text{res}} \right|.$$

This gives for the combined cross-section for inelastic processes

101 **Resonance and potential scattering**

$$\sigma_{\text{inelastic}} = \pi \lambda^2 \frac{-4ka \, \text{Im} f}{|f - ika|^2}. \qquad\qquad 7.4$$

Thus if f is real the interaction is purely elastic, with no reactions occurring; but if reactions are to occur, their cross-section is given by the imaginary part of f, which must be intrinsically negative.

7.2.4. The next step

We are now in a position to look inside the target nucleus and consider how the processes occurring there will affect the observed cross-sections. As we have already seen, it is sufficient to know what effect these processes have on the value of f, the logarithmic derivative of the radial wave function, at the surface of the nucleus.

To evaluate this, we use the principle of continuity, by which the values of the wave function and of its gradient must be continuous across the boundary between the inside and the outside of the nucleus. These two conditions are satisfied automatically if we calculate the value of f from the wave function describing the inside of the nucleus, and equate it to the value calculated in section 7.2 from the assumed wave function outside.

In the full treatment, partial waves with $l \neq 0$ must be considered also. They obey relations similar to those developed above for $l = 0$, with some extra complications.

7.3 Formation of compound nucleus

For the purpose of Chapter 6, and for most discussions of reactions which involve an intermediate state, it is assumed that there are only two possible answers to the question, 'Was a compound nucleus formed?' 'Yes' and 'no' are allowed, but not 'partly'. The dividing line between the two permitted answers is fixed by the boundary of the nucleus, which is assumed to be well-defined. As soon as the incident particle enteres the target nucleus, it is irretrievably absorbed, and the compound nucleus has started to exist.

This immediate and complete absorption is represented by the absence of any outgoing spherical partial waves inside the nucleus. For a single partial wave with $l = 0$, the radial wave function then contains only one term,

$$u(r) = e^{-iKr}, \qquad\qquad 7.5$$

where K is the wave number of one of the incident particles when inside the nucleus. If the nucleus has a potential well of depth V, an incident particle of kinetic energy E will have

$$K = \hbar^{-1}\sqrt{\{2M(V+E)\}}.$$

This may be compared with its wave number outside the nucleus, which is

$$k = \hbar^{-1}\sqrt{(2ME)}.$$

The wave function of equation **7.5** has a logarithmic derivative at the surface nucleus equal to

$$f = -ika.$$

Fitting this to equation **7.1**, we get

$$\eta e^{2i\delta} = \frac{K-k}{K+k}, e^{-2ika},$$

whence $$\eta = \frac{K-k}{K+k}$$

and $$A_{res} = \frac{2k}{K+k} \quad .$$

The combined cross-section for all the inelastic processes, that is, all reactions proceeding via formation of the compound nucleus, is

$$\sigma_{inelastic} = \pi\lambda^2(1-\eta^2)$$

$$= \frac{4\pi K}{k(K+k)}. \tag{7.6}$$

This expression should hold in conditions where the compound nucleus is highly excited, with levels overlapping, so that the incident particle can be absorbed at any energy. The condition that there should be no outgoing spherical waves inside the nucleus also implies that compound elastic scattering is very weak; that is, that the compound nucleus is much more likely to break up in a reaction channel than to re-form elastically the particles from which it was created. Thus equation **7.6** is not expected to be valid at low energies, where it gives incorrect predictions.

At intermediate energies, the hypothesis of complete absorption by a continuum of levels may need to be replaced by a model of absorption through excitation of a single resonance. This is considered in the following section.

7.4 Resonance reactions

If an incident particle is absorbed in such a way that a definite resonant state of a compound nucleus is excited, the wave function inside the nucleus is not given by equation **7.5**. Instead it is a system of spherical waves with properties depending on the shape of the potential well. For a square well, if there is no decay except by the elastic channel (e.g. because the energy is too low for any other channel), the radial wave function has outgoing and ingoing components of equal amplitude, and may be represented by

$$u(r) = e^{i(Kr+2\zeta)} + e^{-iKr}$$

$$= \text{constant} \times \cos(Kr+\zeta),$$

which gives $$f = -Ka\tan(Ka+\zeta),$$

103   Resonance reactions

where the phase shift ζ is a real angle dependent on the energy. At a resonance ζ is such that f is zero (see Blatt and Weisskopf, 1952, p. 381), and near a resonance f is small.

A real value of f, whether small or not, gives purely elastic scattering with no reactions. To allow for the possibility of reactions, it is clearly necessary to include a term representing the decay of the resonance. This introduces an imaginary component of f.

If the real part of f is zero at the resonance energy E_0, we may allow for the energy dependence of K, and include the imaginary term, by using, for an energy E close to the resonance, an expression

$$f = \alpha(E - E_0) - i\beta, \qquad\qquad 7.7$$

where α and β are small real positive constants.

With f given by equation 7.7, A_{res} takes on the value

$$A_{\text{res}} = \frac{-2ika}{\alpha(E - E_0) - i\beta - ika}$$

$$= \frac{-2ika/\alpha}{(E - E_0) - (i/\alpha)(\beta + ka)}.$$

This may be very much larger than A_{pot}. In such cases A_{pot} may be neglected in the calculation of the cross-section for elastic scattering,

$$\sigma_{\text{elastic}} = \pi \lambda^2 \,|\, A_{\text{res}} \,|^2$$

$$= \pi \lambda^2 \frac{k^2 a^2/\alpha^2}{(E - E_0)^2 + \{(\beta + ka)/\alpha\}^2}. \qquad\qquad 7.8$$

The corresponding expression for the cross-section for inelastic processes is

$$\sigma_{\text{inelastic}} = \pi \lambda^2 \left[1 - \left| \frac{f + ika}{f - ika} \right|^2 \right]$$

$$= \pi \lambda^2 \frac{4ka\beta/\alpha^2}{(E - E_0)^2 + \{(\beta + ka)/\alpha\}^2}. \qquad\qquad 7.9$$

Equations 7.8 and 7.9 fit the form of the standard Breit–Wigner equations 6.12 and 6.13, if we write for the partial width of the entrance channel

$$\Gamma_a = \frac{2ka}{\alpha},$$

and for the total level width

$$\Gamma = \frac{2ka}{\alpha} + \frac{2\beta}{\alpha},$$

with $2\beta/\alpha$ representing the sum of the partial widths for all the reaction channels.

7.5 The optical model

Returning to our review of reaction mechanisms in section 7.1, we see that the models used so far have been extreme, including no attempt to describe the early stages of the process by which an incident particle actually loses its kinetic energy.

As a better approximation to a smoothed-out description of what actually happens, it is common to give the nucleus a square well with a complex potential

$$V = -(V_0 + iW).$$

This gives a wave function inside the nucleus with ingoing and outgoing spherical partial waves, of the form

$$u(r) = e^{\pm iKr}.$$

The wave number K is now complex and must be expressed as

$$K = \hbar^{-1}\sqrt{\{2M(E + V_0 + iW)\}}$$

$$= K_R + iK_I.$$

If W is small compared with V_0, K_R and K_I are given by

$$K_R = \hbar^{-1}\sqrt{\{2M(E + V_0)\}}$$

$$\text{and} \quad K_I = \hbar^{-1}W\sqrt{\frac{M}{2(E + V_0)}}$$

$$= \frac{WK_R}{2(E + V_0)}.$$

K_R, the real part of the complex wave number, is just the reciprocal of $\lambdabar$ for the waves representing the motion of the incident particles inside the nucleus. But K_I, the imaginary part of the wave number, is proportional to the imaginary part of the potential and is related to the exponential absorption of these waves. The mean free path for absorption of the incident particles in their passage through the target nucleus is $1/2K_I$. In this context 'absorption' must be interpreted as meaning the occurrence of a nucleon–nucleon collision with transfer of enough energy to prevent either of the products from being subsequently considered as part of the incident beam. In the extreme form of the compound-nucleus model, this mean free path is considered to be infinitely small. But the complex potential of the optical model gives a convenient way of relaxing this often unrealistic condition.

For a given complex potential, a wave function inside the nucleus has to be set up and fitted to the wave function outside, to allow determination of the elasticity η and the phase shift δ in terms of the parameters of the nuclear potential. The first presentation of the optical model, with calculations following these lines, was made by Feshbach, Porter and Weisskopf (1954), who calculated total cross-sections for interaction of neutrons with many target elements. They obtained good general agreement with experimental observations of the way these cross-sections depend on neutron energy and on target mass.

More recent calculations, profiting from the improvement in electronic computers, have been able to include more detailed descriptions of the nuclear potential. For example, Glassgold and Kellogg (1958) made calculations for protons, with the effect of Coulomb forces included, and the perfect square well replaced by one with a diffuse edge. They obtained best fits to experimental results for 40 MeV protons with a complex potential

$$V = (36 + i15) \, \text{MeV},$$

spread over a square well of radius $1 \cdot 30 A^{\frac{1}{3}}$ fm, with surface thickness $0 \cdot 7$ fm.

The name 'optical model' comes from the fact that the predicted angular distributions for elastic scattering show features similar to those of the diffraction patterns obtained when light is scattered by a translucent object.

7.6 Direct reactions

7.6.1 (d, n) *stripping reactions*

In contrast to the reactions which have been discussed in terms of the optical model and of the compound-nucleus model, the present section deals with reactions which involve direct passage from the initial to the final state, without formation of any intermediate state.

The best-known direct reactions are the stripping reactions, in which an incident deuteron is stripped of one of its nucleons, which enters the target nucleus, leaving the other to continue an independent life outside. In the case of a (d, n) reaction, it is the proton that enters the target nucleus, while the neutron remains outside, as shown in Figure 28.

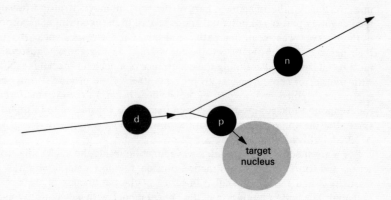

Figure 28 (d, n) stripping reaction

The properties of the particular target nucleus determine the details of the reaction in fixing the energy and the angular momentum with which the proton must enter it. The reaction is not usually observed unless these quantities are correct for forming a well-defined state of the final nucleus.

The orbital angular momentum l which the proton needs to take it into the target nucleus is determined by the angular momenta and parities of the initial and final nuclear states. Conservation of angular momentum limits l to one of two consecutive integral values, since the difference between the total angular momenta of the initial and final states must be equal to $l+\frac{1}{2}$ or $l-\frac{1}{2}$, according to whether the proton has its spin of one half parallel or antiparallel to l. Conservation of parity (see Hughes, 1971) fixes which of the two values of l will be effective, for if the parities of the initial and final nuclei are the same, l must be even, but if the initial and final nuclear states are to have different parities, l must be odd.

The kinetic energy of the neutron is determined by how much of the incident energy is taken into the nucleus by the proton. In fact it serves to identify the final state produced, by the Q-value of the reaction as a whole.

The theory of stripping reactions was first outlined, for low energies and without discussion of angular momentum, by Oppenheimer and Phillips (1935). More detailed treatments, concerned especially with using angular distributions to assign values of angular momentum to nuclear states formed in stripping reactions at higher energies, have been published by Butler (1951) and by Bhatia et al. (1952).

The basic theory, as outlined by Butler and Hittmair (1957) starts with three approximations, namely that the following interactions may be neglected: (a) interaction of the continuing particle – neutron in a (d, n) reaction – with the initial nucleus; (b) interaction of the neutron with the proton after one of them has entered the nucleus; (c) interaction of the target nucleus with the deuteron as a whole, leading to elastic scattering.

Under these approximations, which are valid for incident deuteron energies greater than a few million electronvolts, the problem reduces itself to writing down an initial wave function to describe the target nucleus with an incident plane wave of deuterons, including the internal wave function of the deuteron, and relating this to a final wave function. The latter must describe the captured proton inside the nucleus and the neutron departing in partial waves with l determined by the angular momentum with which the proton entered the nucleus. Capture of the protons with a given l thus causes the neutrons to depart in a definite set of partial waves, and hence with a characteristic angular distribution. For example, capture with $l = 1$ in the reaction $^{11}B(d, n)^{12}C$ at 8 MeV gives a first peak in the angular distribution of neutrons at about $20°$.

While the angular distribution in a stripping reaction is determined largely by angular momentum, the absolute magnitude of the cross-section is determined by the probability that the system consisting of the target nucleus and the nucleon entering it will become a single nucleus in the required final state. This quantity depends on the coupling between the states, which in the converse sense describes the fraction of the time which the final nucleus would normally spend in the state consisting of an unexcited target nucleus plus one nucleon of the right energy. In the language of Chapter 6, the partial width for this particular capture channel determines the cross-section for the stripping process which depends upon the capture.

7.6.2 (d, p) *stripping reactions*

If it is the neutron that is captured, while the proton remains outside the nucleus, the process is a (d, p) reaction which may be described by theory differing only marginally from that outlined above. The chief difference is that allowance must be made for the Coulomb interaction of the departing proton with the final nucleus. This results in no qualitative changes, but the peaks in the angular distributions are shifted outwards to slightly greater angles. If the effect were ignored, it could in some cases lead to mistakes in assignment of angular momentum to the final nucleus.

On the other hand, treatment of (d, p) reactions is simpler in that the inter-action of the captured particle with the target nucleus contains no Coulomb component. Exact handling of this component for a (d, n) reaction is laborious and approximate solutions are usually adopted.

7.6.3 *Pick-up reactions*

With incident energies less than about five million electronvolts (n, d) and (p, d) reactions are likely to proceed mainly through compound-nucleus formation. At higher energies the inverse of a stripping reaction may be important. If the target nucleus has an appreciable probability of being found in a state equivalent to a free neutron and a residual nucleus in a state which could exist alone, a passing proton may pick up the neutron to form a deuteron with it. For this to happen, the neutron must be in a state which allows it, when free, to have momentum close to that of the incident proton; if there is too much discrepancy, the interaction will be too weak to form the deuteron.

While the above is only a qualitative description of a (p, d) reaction occurring by pick-up, it serves to explain the principles of the more exact treatment, which is related to that of the stripping process by the reciprocity theorem (see section 6.3.3). The angular distribution of the final deuterons is determined by the angular momentum of the state from which the neutron was picked up. Thus a pick-up reaction in a nucleus with an odd neutron outside a closed inner shell can provide useful information about the state of the odd neutron, information which is directly relevant to the shell model of the nucleus (see Reid, 1972) and valuable in confirming the sequence in which levels are filled.

Chapter 8
Neutrons

8.1 General properties of neutrons

8.1.1 *The significance of slow neutrons*

Positively charged particles of energy less than 100 keV do not have great interest to the nuclear physicist, because they are very unlikely to get close enough to a nucleus to be within the range of nuclear forces. The limits of significant penetration of the potential barrier at low energies are discussed in Chapter 10. Negatively charged particles of low energy can settle in atomic orbits around nuclei; for electrons these orbits have radii 10^{-8}–10^{-10} cm. When the process is such as to release energy, capture of an electron from one of these orbits may occur as an inverse β-decay process. However, if a negative pion is captured to form a pi-mesic atom, its orbit has a smaller radius because its mass is greater than that of an electron. The normal fate of such a pion is capture by the nucleus, where it starts a nuclear reaction leading to the more or less complete disintegration of the nucleus.

Furthermore, as charged particles pass through matter, they suffer energy loss through ionization: frequent collisions with electrons transfer to the latter energies sufficient to detach them from their parent atoms. The frequency of such collisions varies with the velocity of the moving charged particle. According to the Bragg curve of ionization against residual range, the ionization rises as velocity decreases, reaching a peak at a velocity of order $10^{-2}c$ and thereafter falling as the moving charge is increasingly neutralized by dragging electrons along with it. In general the last hundred thousand electronvolts of kinetic energy of a charged particle passing through matter is lost by ionization in a distance so short that there is very little chance of any nuclear interaction.

On the other hand, electrically neutral particles like neutrons can pass through matter without causing ionization, losing no energy except when they interact with nuclei. These interactions are subject to no potential barrier, and have cross-sections which, instead of dropping steeply with decreasing energy, rather tend towards high values at low energies.

It follows that neutron beams have important nuclear properties over a wide range of energies, which may be subdivided as follows, according to a rough allocation of predominating properties.

8.1.2 Classification of neutrons

Fast neutrons. This term may be used to describe neutrons as they emerge from nuclear reactions, with kinetic energies typically between 1 and 20 MeV.

Intermediate neutrons. These neutrons have energies between 1 MeV and 1 keV.

Resonance neutrons. Neutrons with kinetic energies between 1 keV and 1 eV, are so called because this is the interval in which there is the most striking resonance absorption, through excitation of target nuclei to well-defined excited states.

Epithermal neutrons. These neutrons, with energies from 1 eV to about 0·05 eV, are neutrons whose energies have fallen nearly to thermal values.

Thermal neutrons. Properly speaking, these are neutrons which, through repeated collision with nuclei of material which is macroscopically at rest, have acquired energies in statistical equilibrium with the thermal energies of these nuclei at the particular temperature of the material. In practice the term is usually taken to imply energies which would be in equilibrium with those of nuclei in material at a normal room temperature of about 20°C. In these conditions, the neutrons have a Maxwellian distribution of velocities, the peak being at an energy $k\Theta = 0·025$ eV.

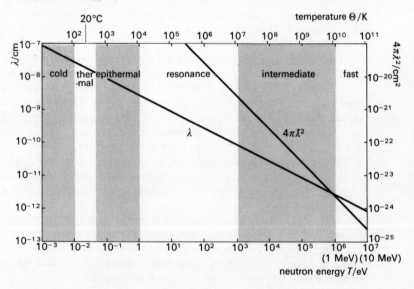

Figure 29 Relation between neutron energy T, wavelength λ, maximum possible cross-section ($= 4\pi\lambda^2$) and temperature Θ giving Maxwellian peak at $k\Theta = T$. All scales logarithmic

Cold neutrons. The name implies that these might be neutrons in thermal equilibrium with material at low temperature. So far as energy is concerned this is correct, but in fact they are usually obtained by using a velocity selector to separate out the slower components of a beam of ordinary thermal neutrons. 'Cold' is a loose term, referring to kinetic energies of order 10^{-3}eV, which would be in thermal equilibrium with material at a temperature of 12 K.

The broader term 'slow neutrons' is often convenient for covering the whole range from about 1 keV downwards.

The relation between temperature Θ and peak kinetic energy $T = k\Theta$ is shown in Figure 29, along with the corresponding values of λ and $4\pi\lambda^2$ (see section 8.1.3).

8.1.3 Wavelength

The nature of the interaction between a beam of neutrons and the nuclei of a target may be discussed in terms of the de Broglie wavelength of the neutrons. This is

$$\lambda = \frac{h}{p} = \frac{h}{\sqrt{(2mT)}}$$

for neutrons of momentum p, kinetic energy T and rest mass m, when

$T \ll mc^2$.

If we insert numerical values, expressing T in electronvolts, we get

$\lambda = 2{\cdot}86 \times 10^{-9} T^{-\frac{1}{2}}$ cm,

an expression which yields the values shown in Table 8.1, and in Figure 29.

Table 8 Wavelength of Neutrons as a Function of Kinetic Energy

T/eV 10^6	10^3	1	10^{-2}	10^{-3}
λ/cm $2{\cdot}86 \times 10^{-12}$	$9{\cdot}04 \times 10^{-11}$	$2{\cdot}86 \times 10^{-9}$	$2{\cdot}86 \times 10^{-8}$	$9{\cdot}04 \times 10^{-8}$

As we have seen in Chapters 5 and 6, cross-sections for nuclear processes may be considered in terms of a maximum possible total cross-section which is $4\pi\lambda^2(= \lambda^2/\pi)$. This increases as we go to lower and lower energies, reaching 10^{-24} cm^2 at $T = 2{\cdot}6$ MeV, and 10^{-20} cm^2 at $T = 260$ eV (see Figure 29). Cross-sections approaching these limits are not unknown, though it should be emphasized that normal non-resonant values are much smaller.

However, the wavelength associated with a beam of neutrons is important in another sense. When the target is a solid, with nuclei arranged in a regular lattice with characteristic interatomic spacings of order 10^{-8} cm, the elastic scattering of neutrons with wavelengths also of this order will be coherent. Instead of taking each target nucleus as scattering independently with appropriate cross-section, one has to sum the amplitudes of all the scattered waves in a particular direction, squaring the result to get the intensity, just as one does when considering the diffraction of X-rays by a crystal. In these conditions the concept of cross-section ceases to be helpful, and interest is transferred to the interference phenomena which lead to the use of neutron diffraction as a tool for the study of crystal structures. While X-ray diffraction patterns reveal the spatial distribution of electron density in the target material, neutron diffraction reveals the distribution of nuclear matter, and hence the distribution of mass. This difference can be significant, for example when the position of a hydrogen atom in an organic molecule is in question.

8.2   Neutron sources

8.2.1   Radioactive sources

Fast neutrons may be obtained from the (α, n) reactions in light elements such as lithium, beryllium, boron and fluorine. With the α-particles from radioactive materials, natural or man-made, beryllium gives a neutron spectrum extending up to 11 MeV. Lithium, boron and fluorine give lower yields, with somewhat lower neutron energies. They therefore tend to be used only when the details of the neutron spectrum are important; for example, in simulating the spectrum of neutrons from fission of uranium.

The possible α-emitting radioactive elements include americium-241, polonium-210 and radium-226. Polonium-210 was the first to be used (see section 4.3.1), and can now be obtained in very intense sources (e.g. 50 curies); but its mean life of 138·4 days is inconveniently short for many purposes. Also, groups of polonium atoms are liable to become detached from a solid source by evaporation or recoil; if these are carried off by the air, they may constitute a serious danger to health. The man-made element americium-241, with mean life 458 years, is good for sources up to about 5 curies, but they cost nearly as much as polonium-210 sources which are ten times more intense.

The radioactive material is usually made up with the target element into a pellet which is kept and used in a stainless steel cylinder. Yields range from about 2×10^3 neutrons per second (from one millicurie ^{241}Am–Be, cost £45) to 10^8 neutrons per second (from 50 curies ^{210}Po–Be, cost £1650) (prices from 1969 catalogue of Radiochemical Centre, Amersham).

8.2.2   Electrostatic neutron generators

Increasingly important nowadays are small electrostatic machines which accelerate deuterons to an energy of order 100 keV, enough to produce 14 MeV neutrons from a tritium target by the D + T reaction (see section 4.3.2)

$$^2_1H + {}^3_1H \longrightarrow {}^1_0n + {}^4_2He \quad (Q = 17·577 \text{ MeV}).$$

It is possible (see Figure 30) to make a sealed beam-and-target tube containing deuterium and tritium gas at low pressure, with a target of mixed deuteride

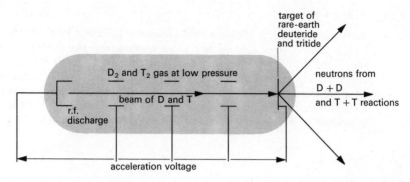

Figure 30 Sealed beam-and-target tube for neutron source (diagrammatic)

and tritide of a suitable rare-earth element. A built-in radio-frequency ion source produces deuterons and tritons which are accelerated towards the target by a potential difference of about 100 kV from an external electrostatic generator. In this way any loss of deuterium and tritium from the target is made up by the ions from the gas, so the life of the tube is greater than that of a simple target.

The energy released in the D + T reaction, 17·577 MeV, is so great that the neutron energy is effectively the same (14·3 MeV) whether the deuteron or the triton is the target. There is of course a contamination of 3–4 MeV neutrons from the D + D reaction, but this is not a serious disadvantage of the mixed system since even a target which is meant to be a pure tritide contains some deuterium after a period of bombardment by deuterons. There are also contributions from the T + T reaction

$$_1^3\mathrm{H} + {}_1^3\mathrm{H} \longrightarrow {}_0^1\mathrm{n} + {}_2^5\mathrm{He} \quad (Q = 10\cdot37 \text{ MeV})$$

and from the spontaneous decay of ^{5}He which follows it

$$_2^5\mathrm{He} \longrightarrow {}_0^1\mathrm{n} + {}_2^4\mathrm{He} \quad (Q = 0\cdot95 \text{ MeV}).$$

These machines can be bought as self-contained units capable of producing continuous or pulsed neutron yields up to about 10^{11} neutrons per second. The spectrum is predominantly the D + T peak at 14·3 MeV, the same in all directions.

Some variation of energy with direction may be obtained by using a higher bombarding energy, but the D + T reaction has such a large Q-value that this effect is of very limited use in giving a variable-energy neutron source. For example, even with 2 MeV incident deuteron energy, the total range of neutron energies, from 0° to 180°, is only from 18·0 MeV to 12·2 MeV. If the tritons are incident with 2 MeV kinetic energy, on stationary deuterons, the range of neutron energies available is from 18·7 MeV to 11·2 MeV. For the range of energies below 12 MeV, other reactions with smaller Q-values are more convenient.

8.2.3 Reactors

Slow neutrons may be obtained from a source of fast neutrons by interposing a thick layer of some light element to act as moderator (see section 8.3), and this is the only practicable source of slow neutrons in many laboratories. However, slow neutron beams many orders of magnitude more intense can be obtained from reactors in which fission of heavy elements is proceeding by a chain reaction. This is a specialized topic, which forms the subject of Chapter 9, so it will be developed no further under the present heading.

8.3 Moderation

8.3.1 Loss of energy by fast neutrons

Elastic scattering. As has been mentioned already, neutrons can pass through matter without losing energy through the electrical interactions which dominate the behaviour of charged particles. The only processes by which neutrons lose energy are nuclear interactions with stationary nuclei. Most of these interactions lead to elastic scattering, in which the neutron may lose energy up to a limit set by the ratio of its mass to that of the struck nucleus.

Equation **3.35** showed that the kinetic energy transferred from a non-relativistic incident particle of mass m to a previously stationary particle of mass M could range from zero to a fraction $4Mm/(m+M)^2$ of the incident kinetic energy. The minimum possible kinetic energy of a neutron of mass m, after collision with a nucleus of mass M, is thus a fraction

$$a = \left(\frac{M-m}{M+m}\right)^2 \quad \text{(see equation 3.37)}$$

of its initial kinetic energy T_0.

The actual kinetic energy T_1 remaining after scattering at a c.m.s. angle θ_c is given by equation **3.36**, which in the notation of the present discussion becomes

$$T_1 = \tfrac{1}{2}\{1+a)+(1-a)\cos\theta_c\}\, T_0.$$

Logarithmic decrement. The fractional change of kinetic energy may conveniently be measured on a logarithmic scale, so that the mean logarithmic change of kinetic energy in a series of collisions can be expressed as the sum of the mean logarithmic energy changes for the individual collisions.

If we start with

$$\ln\frac{T_1}{T_0} = \ln\left[\frac{(1+a)+(1-a)\cos\theta_c}{2}\right],$$

we may calculate the mean value of $\ln(T_1/T_0)$ by assuming the scattering to be isotropic in the c.m.s. This assumption is valid for S-wave scattering, which is the only type having any significance at the energies to be considered in the present chapter. It gives a uniform number of scattered neutrons per unit c.m.s. solid angle, and therefore per unit interval of $\cos \theta_c$. Observed values of $\cos \theta_c$ are therefore distributed uniformly over the range from 1 to -1, and the mean value of $-\ln(T_1/T_0)$, which is known as the logarithmic decrement ξ, is found by integrating with respect to $\cos \theta_c$ over this interval, as follows:

$$\xi = -\overline{\ln\frac{T_1}{T_0}}$$

$$= \frac{-\int_{-1}^{1} \ln \tfrac{1}{2}\{(1+a)+(1-a)\cos \theta_c\} \, d(\cos \theta_c)}{\int_{-1}^{1} d(\cos \theta_c)}$$

$$= 1+\frac{a \ln a}{(1-a)}. \qquad\qquad \textbf{8.1}$$

The mean value of $\ln(T_0/T_n)$, the logarithmic energy change in a series of n collisions, is therefore

$$-\ln \frac{T_n}{T_0} = n\xi = n\left[1+\frac{a \ln a}{1-a}\right].$$

The approach to thermal energy. The energy of a neutron falls off exponentially as it makes successive collisions. In the end, when the kinetic energy becomes not much greater than the energies of thermal motion of the target nuclei, it approaches exponentially towards a thermal limit of $\tfrac{3}{2}k\Theta$ instead of towards zero. In fact the velocities of thermal neutrons in a material at temperature Θ are distributed according to the Maxwell–Boltzmann formula, and it has become common to speak of them as having energy $k\Theta$. This is actually the energy corresponding to the peak of the velocity distribution; that is, it is the most probable energy per unit velocity. The true mean energy is $\tfrac{3}{2}k\Theta$.

This process of loss of energy and approach to thermal energy is known as moderation, and a substance in which it occurs is called a moderator.

Equation **8.1** shows that the efficiency of a moderator, per collision, may be measured by the logarithmic decrement ξ, which is tabulated for several light elements in Table 9.

Logarithmic decrement for hydrogen. For hydrogen, with target mass and neutron mass effectively equal, $a = 1$ and the expression **8.1** for ξ is indeterminate. However, it may be shown to tend, as a approaches 1, towards a limit

$$\xi_{\mathrm{H}} = 1.$$

Alternatively the same result may be obtained more simply from the kinematics of the n + p collision, from which a scattered neutron has energy

$$T_1 = T_0 \cos \theta = T_0 \tfrac{1}{2}(1+\cos \theta_c)$$

for a c.m.s. angle θ_c and a laboratory-system angle θ. The value of T_1 ranges from T_0 at $\theta = 0°$ to zero at $\theta = 180°$, with a mean logarithmic value given by

$$\xi_H = -\overline{\ln \frac{T_1}{T_0}}$$

$$= -\tfrac{1}{2} \int_{-1}^{1} \ln \tfrac{1}{2}(1 + \cos \theta_c) \, d(\cos \theta_c)$$

$$= 1.$$

Thus hydrogen is the best moderator of all, from the point of view of energy loss per collision. However, later sections will reveal that there are other qualifications for a good moderator.

Table 9 Parameters of Neutron Collisions in Light Elements

Element	H	D	^{12}C	^{16}O
A	1	2	12	16
Minimum fraction of energy remaining after collision, a	0	0·111	0·716	0·78
Logarithmic decrement, ξ	1	0·725	0·155	0·120
Mean number of collisions for thermalization $\bar{n} = 18\cdot2/\xi$	18·2	25·0	117	152
$\overline{\cos \theta} = b = 2/3A$	0·667	0·333	0·056	0·042

Number of collisions for thermalization. The average number of collisions needed to reduce the kinetic energy of a neutron from a typical 'fast' value T_0 to a typical thermal value T' may be taken, in first approximation, as

$$\bar{n} = \frac{\ln(T_0/T')}{\xi}. \qquad \textbf{8.2}$$

For listing the values of $\bar{n}$ in different moderators, it has become conventional to choose a value of 2 MeV for T_0, and for T' to use 0·025 eV, which is very close to the value of $k\Theta$ at a temperature of 20°C. The ratio of these gives

$$\bar{n} = \frac{18\cdot2}{\xi}. \qquad \textbf{8.3}$$

In accordance with convention, this expression has been used for calculating the values of $\bar{n}$ listed in Table 9. However, two further points need to be made before these figures are given too precise a quantitative significance:

(a) The final exponential approach is toward a mean thermal energy T_{th}, not towards zero; $\bar{n}$ collisions, with $\bar{n}$ specified by equation **8.2**, will therefore actually give a logarithmic mean energy $T = T' + T_{th}$.

(b) The true value of T_{th} is $\frac{3}{2} k\Theta$, not $k\Theta$.

It follows from the combined effect of these two points that values of $\bar{n}$ such as those in Table 9 really refer to moderation from 2 MeV to a mean energy about $\frac{5}{2} k\Theta$, or alternatively from 3 MeV to 3 kΘ, or from 1 MeV to 2 kΘ.

Slowing-down power. As a neutron moves through a moderating material, its probability of scattering, per unit distance travelled, is given by $N\sigma$, where N is the number of moderator nuclei per unit volume, and σ is their cross-section for elastic scattering of neutrons.

The mean logarithmic decrease of kinetic energy per unit distance travelled by the neutrons is therefore

$$S = \xi N\sigma.$$

This quantity is known as the slowing-down power; it has the dimensions length^{-1}, and it measures the efficiency of a moderating material, as the average logarithmic decrease of kinetic energy per unit length of path of a neutron. The average energy loss per unit length of neutron path is not easy to calculate, since average logarithmic decrease is not the same as log(average decrease), but one may think of e^{-S} as the factor by which the kinetic energy of an unmoderated neutron is typically reduced in unit path length. Thus S is of slightly more practical significance than ξ itself, which describes the probable effect of a collision without giving any indication of how likely a collision is to occur.

In a material such as water, which contains more than one type of nucleus, the total probability of scattering per unit path is $\Sigma N\sigma$, summed over all the types of nucleus present.

Similarly the slowing-down power of such a material is given by

$$S = \sum \xi N\sigma,$$

each element contributing in proportion to its logarithmic decrement per collision, its number of nuclei per unit volume and its cross-section for elastic scattering of neutrons. Values for a few important moderators are listed in Table 10.

Moderating ratio. At this point it is necessary to mention that neutrons can suffer nuclear reactions as well as elastic scattering. If a moderator is intended to slow down fast neutrons to thermal energies, any absorption of neutrons by capture or other reactions will reduce its suitability for the purpose. A parameter which gives a measure of the fraction of the neutrons which are slowed down without being absorbed is the moderating ratio. This is obtained by multiplying the slowing-down power by λ_a, the mean free path for absorption. This λ_a is the reciprocal of $\Sigma\, N\sigma_{abs}$, which is the total probability of absorption per unit distance covered by a neutron. As in the calculation of the slowing-down power, the sum is taken over all the types of nucleus present, each contributing in proportion to its cross-section for absorption of neutrons, and to the number N of such nuclei per unit volume.

Table 10 Properties of Moderators

Material	H_2O	D_2O	C (graphite)
Mean free paths for thermal neutrons/cm			
scattering λ	0·37	2·2	2·6
transport λ_t	0·45	2·6	2·75
absorption λ_a	49	30000	3000
Diffusion constant $D = \frac{1}{3}\lambda_t$/cm	0·15	0·87	0·92
Diffusion length $L = \sqrt{(\frac{1}{3}\lambda_t\,\lambda_a)}$/cm	2·7	178	51
Slowing-down power $\times$ cm	1·36	0·17	0·063
Moderating ratio	62	5000	190
Age for thermalization τ_{th}/cm^2	32	120	350
Slowing-down length $= \sqrt{\tau_{th}}$/cm	5·7	11·0	18·7

The moderating ratio, defined in this way, is thus the mean total logarithmic loss of energy per neutron lost by absorption; that is, the mean number of factors e of energy loss per neutron absorbed. Values for three important materials are listed in Table 10. The relatively low value for water results from radiative capture of neutrons by protons, according to

$$^{1}_{1}\mathrm{H} + {}^{1}_{0}\mathrm{n} \longrightarrow {}^{2}_{1}\mathrm{H} + \gamma.$$

This reaction has a low cross-section, but it is enough to remove about one neutron for every three brought to thermal energies by moderation in water. Deuterium on the other hand absorbs so few neutrons that heavy water has a moderating ratio of 5000, representing an average of one neutron lost per 270 successfully moderated.

8.3.2 *The distance between collisions*

Mean free path. To obtain from a number of collisions, for example, $\bar{n}$ the mean number required to bring a neutron to thermal energy, some measure of the thickness of moderator required, we have to insert the mean distance travelled between successive collisions. The mean free path, as this quantity is called, is given by

$$\lambda = (N\sigma)^{-1}. \qquad \textbf{8.4}$$

If a material contains several types of nucleus, the over-all mean free path is the reciprocal of the total probability of scattering per unit path,

i.e. $\quad \lambda = \left(\sum N\sigma\right)^{-1}.$

Root-mean-square free path. For some calculations we need, not the mean distance, but the mean square or the root-mean-square (r.m.s.) distance between collisions. These are calculated as follows, the method being given in detail because it is applicable also to paths of molecules in a gas, to time intervals between randomly occurring events, or to any type of interval between processes which occur randomly with uniform probability.

In the present case the probability of collision per unit length is $N\sigma$, so the probability of collision in an element of path ds is $N\sigma\,ds$ per neutron reaching the element.

Let $\pi(s)$ be the probability that a neutron covers a distance s from a collision without suffering further collision; then the probability of collision in the element of path between s and $s+ds$ is $N\sigma\,ds\,\pi(s)$. This must also be equal to $-\pi'(s)\,ds$, whence we conclude that

$$\pi'(s) = -N\sigma\,\pi(s)$$

and $\quad \pi(s) = e^{-N\sigma s}.$

The over-all probability that a neutron will survive a distance s without collision and then be scattered in the next element ds is therefore

$$P(s)\,ds = -\pi'(s)\,ds = e^{-N\sigma s}.\,N\sigma\,ds.$$

The mean free path λ is then calculated by putting this function into

$$\lambda = \bar{s} = \frac{\displaystyle\int_0^\infty s\,P(s)\,ds}{\displaystyle\int_0^\infty P(s)\,ds},$$

which integrates to $(N\sigma)^{-1}$, in agreement with **8.4**.

This technique allows us to calculate the mean-square path length by integrating as follows:

$$\overline{s^2} = \frac{\int_0^\infty s^2 \, P(s) \, ds}{\int_0^\infty P(s) \, ds}. \quad = 2(N\sigma)^{-2} = 2\lambda^2. \qquad 8.5$$

The r.m.s. distance between collisions is therefore $\sqrt{2}\lambda$.

Total distance for n collisions in a heavy material. The total distance travelled by a neutron in slowing down has a mean value of order $\bar{n}\lambda$. But this distance is covered in $\bar{n}$ straight lines of mean length λ, more or less randomly directed. The mean distance from the starting point to the last collision is therefore much less than $\bar{n}\lambda$.

For a material with infinitely heavy nuclei (which would not moderate at all) the distance covered in n collisions is easy to calculate, because the centre-of-mass system is the same as the laboratory system, and collisions which are isotropic in the c.m.s. are also isotropic in the laboratory. In this situation the direction of the neutron after a collision is completely uncorrelated with its direction before, and the problem reduces to a random walk of n steps of mean length λ, in completely random directions.

If the steps are actually vectors $\mathbf{s}_1, \mathbf{s}_2, ..., \mathbf{s}_n$, the start-to-finish distance r is given by

$$r^2 = \sum_{i=1}^{n} \mathbf{s}_i \cdot \sum_{i=1}^{n} \mathbf{s}_j$$

$$= \sum_{i=1}^{n} s_i^2 + \sum_{i \neq j} s_i s_j \cos \theta_{ij}. \qquad 8.6$$

When the mean value of r^2 is calculated, the cross-terms vanish because the mean cosine of the angles θ_{ij} between uncorrelated directions is zero. The result is therefore simply

$$\overline{r^2} = n\overline{s^2} = 2n\lambda^2,$$

from **8.5**.

The distance r may be in any direction, and the mean-square component of r along any axis, provided the source is not directional, is

$$\overline{z^2} = \tfrac{2}{3}n\lambda^2. \qquad 8.7$$

The r.m.s. value of the distance covered, and of its component along any axis, are

$$(\overline{r^2})^{\frac{1}{2}} = \sqrt{(2n)}\lambda,$$

$$(\overline{z^2})^{\frac{1}{2}} = \sqrt{(\tfrac{2}{3}n)}\lambda.$$

8.3.3 The distance covered in slowing down

The mean value of cos θ. The basis laid in section 8.3.2 now has to be extended to cover collisions in materials which are useful as moderators. In particular we have to face the fact that if the moderator nuclei are light enough to allow the neutrons to lose some energy in collisions, they are also light enough to spoil the isotropy of the scattered neutrons. In the laboratory system these will in fact tend to move forwards.

To put this in other words, the direction of a scattered neutron will be correlated with its direction before the scattering took place. The angle between these directions is θ, the laboratory-system angle of scattering, which is given for an individual neutron, in terms of the c.m.s. angle $θ_c$, by equation **3.34** (see Figure 8a, p. 43). The extent of the correlation is conveniently specified by the mean value of cos θ, which is given for collisions distributed isotropically in the c.m.s. by

$$\overline{\cos θ} = \tfrac{1}{2} \int\limits_{-1}^{1} \cos θ \, d(\cos θ_c).$$

On substitution from equation **3.34**, this integrates to give

$$b = \overline{\cos θ} = \frac{2}{3A},\qquad\qquad \textbf{8.8}$$

where A is the mass number of the moderator atoms, which is effectively equal to M/m.

Total distance for n collisions in moderator. To calculate the total distance covered in n paths of mean length $λ$, in directions which are correlated to an extent given by a non-zero value of the parameter b, we may use equation **8.6**, which was fuller than was really needed for the simple case it was used to treat.

The cross-terms no longer vanish when the average is taken, and we must write

$$\overline{r^2} = n\overline{s^2} + \overline{\sum_{i \neq j} s_i s_j \cos θ_{ij}}$$

$$= n\overline{s^2} + \overline{s}^2 \sum_{i \neq j} \overline{\cos θ_{ij}}$$

$$= λ^2(2n + \sum_{i \neq j} \overline{\cos θ_{ij}}),$$

since $\overline{s} = λ$ and $\overline{s^2} = 2λ^2$.

The angles $θ_{ij}$ are the angles between all the n paths (see Figure 31), each physical angle appearing twice in the sum. The angles in the sum may conveniently be broken up into different categories as follows. There are $2(n-1)$ angles between adjacent paths, and they have

$$\overline{\cos θ_{ij}} = \overline{\cos θ_{i,i \pm 1}} = b.$$

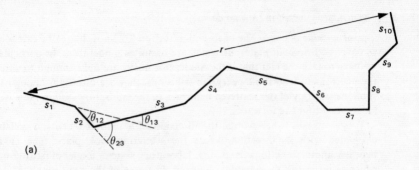

(a)

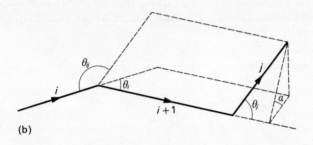

(b)

Figure 31 Motion of neutron in moderator. (a) Distance travelled in n paths with correlated directions. (b) Paths with $j = i+2$

There are $2(n-2)$ angles between paths with $j = i\pm2$. Each of these has a value given by elementary trigonometry as

$$\cos \theta_{ij} = \cos \theta_i \cos \theta_j + \sin \theta_i \sin \theta_j \cos \alpha, \qquad \textbf{8.9}$$

where θ_i is short-hand for the angle $\theta_{i,i\pm1}$ between path i and path $i\pm1$, θ_j for the angle $\theta_{i\pm1,j}$ between path $i\pm1$ and path j (see Figure 31b). The angle α is the azimuthal angle between the plane of paths i and $i\pm1$ and the plane of paths $i\pm1$ and j.

The mean value of $\cos \alpha$ is zero, therefore paths with $j = i\pm2$ have

$$\overline{\cos \theta_{ij}} = \overline{\cos \theta_i} \, \overline{\cos \theta_j} = b^2.$$

Then there are $2(n-3)$ angles between paths with $j = i\pm3$. Each of these can be calculated by extending the result just obtained for paths with $j = i\pm2$. If the mean value of $\cos \theta_{i,i\pm2}$ is b^2, then equation **8.9** with $\theta_{i,i\pm2}$ instead of θ_i, and $\theta_{i\pm2,j}$ for θ_j, gives

$$\overline{\cos \theta_{i,\,j(j=i\pm3)}} = \overline{\cos \theta_{i,\,i\pm2}} \cdot \overline{\cos \theta_{i\pm2,\,j}}$$

$$= b^2 \cdot b = b^3.$$

Continuing up, we find $2(n-4)$ angles with mean cosine b^4, and so on. The result for $\overline{r^2}$ is thus

$$\overline{r^2} = \lambda^2\{2n + 2(n-1)b + 2(n-2)b^2 + 2(n-3)b^3 + \ldots\},$$

which for large n becomes

$$\overline{r^2} = 2n\lambda^2(1 + b + b^2 + b^3 + \ldots)$$

$$= \frac{2n\lambda^2}{1-b}.$$

Thus for a large number of collisions, the mean-square distance covered may be written

$$\overline{r^2} = 2n\lambda\lambda_t, \tag{8.10}$$

where λ_t, which is called the transport mean free path, is given by

$$\lambda_t = \frac{\lambda}{1-b} = \frac{\lambda}{1 - \overline{\cos\theta}}.$$

Finally, we may take a step corresponding to that which gave us equation **8.7**, and say that the general expression for the mean-square component of distance travelled along any axis (see Figure 32) is

$$\overline{z^2} = \tfrac{2}{3}n\lambda\lambda_t. \tag{8.11}$$

Age and slowing-down length. Following an idea introduced by Fermi, we often use the term *age* to describe the stage which neutrons have reached in a process of moderation from the energy at which they were produced. As usually defined, the age τ is one-sixth of the mean-square distance travelled. After n collisions, this is given by equation **8.10** as

$$\tau = \tfrac{1}{6}\overline{r^2} = \tfrac{1}{3}n\lambda\lambda_t. \tag{8.12}$$

The age τ_{th} needed for moderation to thermal energy is obtained by inserting for n the value $\bar{n}$ given by equation **8.3**; thus

$$\tau_{th} = \tfrac{1}{3}\bar{n}\lambda\lambda_t. \tag{8.13}$$

The quantity usually called the slowing-down length is $\sqrt{\tau_{th}}$,

i.e.   Slowing-down length $= \sqrt{\tau_{th}} = \sqrt{(\tfrac{1}{3}\bar{n}\lambda\lambda_t)}$.

This is $\sqrt{\tfrac{1}{2}}$ of the r.m.s. component along each of the three axes of the distance between the point of origin of a neutron and the point at which it reaches thermal energy (see Figure 32).

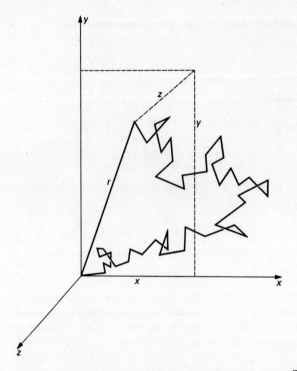

Figure 32 Distance covered by a neutron in many paths. $\overline{r^2}$ is the mean-square distance covered. $\overline{z^2}$ is the mean-square component of distance travelled along one axis, equal to $\frac{1}{3}\overline{r^2}$. When the distance gives moderation to thermal energy $\tau = \frac{1}{6}\overline{r^2}$ = age for thermalization.
$\sqrt{\tau}$ is the slowing-down length, equal to $1/\sqrt{2}$ of r.m.s. z.

Values of the slowing-down length and of τ_{th} are listed in Table 9 for three important materials.

8.3.4 *Diffusion of neutrons*

Transport of thermal neutrons. After neutrons have been brought to thermal energies by the process of moderation, they continue in random motion from collision to collision, with mean energy remaining constant at $\frac{3}{2}k\Theta$. All the calculations of sections 8.3.2 and 8.3.3 for such things as the mean value of $\cos\theta$ and the mean-square distance travelled in a given number of collisions apply to the state of constant energy just as they do when the energy is decreasing at a rate given by the logarithmic decrement.

The process of transport at constant energy is, however, generally called diffusion, since the basic theory is similar to that of the diffusion of gases. In the context of neutron transport this theory may be summarized as follows:

Mean-square distance travelled per unit time. Equation **8.10** gave the mean-square distance travelled in n collisions as $2n\lambda\lambda_t$. This mean time taken for these n collisions, if the neutrons are thermal, is

$$t = \frac{n\lambda}{v},$$

where v is the velocity corresponding to the kinetic energy $\frac{3}{2}k\Theta$. The mean-square distance travelled in unit time is therefore given by substituting v/λ for n, giving

$$\overline{r^2} = 2\lambda_t v.$$

It follows that the mean-square component of distance travelled along any axis in unit time is

$$\overline{z^2} = \tfrac{2}{3}\lambda_t v. \qquad\qquad\qquad \textbf{8.14}$$

Thus although we need the product $2\lambda\lambda_t$ for the mean-square distance travelled per collision, λ_t appears alone with the velocity in the expression for the mean-square distance travelled per unit time.

It should be noted that since the mean-square distance, not the mean or the r.m.s. distance, is proportional to time, there is no single-valued net velocity of transport of neutrons from place to place.

The equation of diffusion. In a uniform volume distribution of neutrons, we may divide the total density into fractions $f_1, f_2, \dots$ (with $\Sigma f = 1$), such that in unit time their displacements along the z-axis are $s_1, s_2, \dots$. When the total density N varies with z, these fractions still apply, and we may calculate the total number of neutrons per unit time crossing the plane $z = 0$ as follows.

Consider the fraction f of neutrons which achieve a vertical displacement s in unit time, half of them upwards and half of them downwards, as shown in Figure 33. If there are $\text{N}(z)$ neutrons per unit volume at z, the number crossing unit area of the plane $z = 0$ per unit time, after starting from points with z-coordinates between z and $z + dz$ ($z < s$) is

$$\tfrac{1}{2}\text{N}(z)f\,dz.$$

These will cross in the direction of decreasing z, and in the same time the number crossing in the direction of increasing z, having started from points between $-z$ and $-(z + dz)$ is

$$\tfrac{1}{2}\text{N}(-z)f\,dz.$$

The net number of neutrons crossing unit area of the plane $z = 0$ per unit time, in the direction of increasing z is

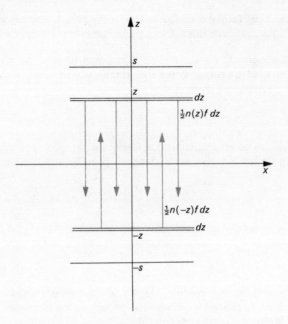

Figure 33 Diffusion of neutrons

$$\tfrac{1}{2}\mathrm{N}(-z)f\,dz - \tfrac{1}{2}\mathrm{N}(z)f\,dz,$$

which, if there are no discontinuities in the distribution, may be written in terms of $\partial\mathrm{N}/\partial z$ as

$$-z\frac{\partial\mathrm{N}}{\partial z}f\,dz.$$

This is the number which cross after having started in one of two layers of thickness dz, at coordinates $\pm z$. Since we are considering neutrons which travel vertical distances $\pm s$, the total number crossing, from all starting points, is

$$\int_0^s -z\frac{\partial\mathrm{N}}{\partial z}f\,dz = -\tfrac{1}{2}\frac{\partial\mathrm{N}}{\partial z}fs^2. \qquad\qquad 8.15$$

Now we can consider neutrons which achieve all possible z-displacements by summing expression 8.15 to give the total number of neutrons crossing unit area of the plane $z = 0$ in the sense of increasing z, from all distances, as

$$-\tfrac{1}{2}\frac{\partial\mathrm{N}}{\partial z}\overline{s^2},$$

where N is now the total number of neutrons per unit volume, and $\overline{s^2}$ is the mean-square vertical displacement per unit time, which is given by equation **8.14** as $\frac{2}{3}\lambda_t v$.

The net rate of transport of neutrons, per unit area per unit time, is therefore given by

$$J = -\tfrac{1}{3}\lambda_t\, v\frac{\partial N}{\partial z}. \qquad\qquad \textbf{8.16}$$

This already takes the form of the standard one-dimensional law of diffusion in which the number crossing unit area per unit time is given by the gradient of number density, with a constant of proportionality known as the diffusion constant.

But conventional neutron diffusion equations are expressed in terms of the neutron flux

$$\phi = Nv$$

rather than of the number density N itself. In terms of this parameter, equation **8.16** becomes

$$J = -D\frac{\partial \phi}{\partial z},$$

where $D = \tfrac{1}{3}\lambda_t$
is the diffusion constant.

This generalizes to three dimensions to give a vector rate of transport of neutrons

$$\mathbf{J} = -D\,\mathrm{grad}\ \phi. \qquad\qquad \textbf{8.17}$$

The rate of increase of number of neutrons per unit volume is given by $-\mathrm{div}\ \mathbf{J}$, so equation **8.17** leads to the second-order diffusion equation for a system in which there is transport of neutrons without either absorption or production

$$D\,\nabla^2\phi = \frac{\partial N}{\partial t}. \qquad\qquad \textbf{8.18}$$

Diffusion in the presence of absorption. The complete diffusion equation must contain a term $-\phi/\lambda_a$ for the rate of absorption of neutrons per unit volume, and a term $P(x, y, z)$ for the rate of production of neutrons per unit volume, for example by fission reactions in which new neutrons are emitted. P will in general be a function of ϕ, and hence of position. The full equation now reads

$$\frac{\lambda_t}{3}\nabla^2\phi - \frac{1}{\lambda_a}\phi + P = \frac{\partial N}{\partial t}. \qquad\qquad \textbf{8.19}$$

When a steady state has been reached, this gives for regions where the source term is zero

$$\nabla^2\phi = \frac{3}{\lambda_\mathrm{t}\lambda_\mathrm{a}}\phi. \tag{8.20}$$

The solutions of this equation for one-dimensional geometry are fluxes decaying exponentially with distance, according to

$$\phi = \phi_0\,e^{-x/L},$$

where L is the diffusion length, given by

$$L = \sqrt{(\tfrac{1}{3}\lambda_\mathrm{t}\,\lambda_\mathrm{a})}.$$

It should be noted that convention has led to a situation in which D, the characteristic distance describing the diffusion process, is called the diffusion constant, while the quantity known as diffusion length is L, which measures the extent to which diffusion escapes being stopped by absorption.

Quantitatively, the diffusion length L is the distance required for decay of the neutron flux by a factor e; this is the same as the mean component of distance diffused along one axis before absorption.

Value of the diffusion lengths are listed, along with other important parameters, for several moderating materials, in Table 10 (p. 118).

8.4 Reactions of neutrons

8.4.1 *Reactions of fast neutrons*

Channels available. When a fast neutron enters a target nucleus, the resulting compound nucleus is usually in a state of very high excitation, from the binding energy of the neutron (of order 8 MeV) plus whatever kinetic energy the neutron has brought in. It is thus likely to undergo rapid de-excitation by one of the following channels:

(a) The elastic channel, in which the neutron is re-emitted and the target nucleus re-formed in more or less its original state. This channel is always available, and it leads to compound elastic scattering which is often an important contribution to the total observed elastic scattering.

(b) Inelastic scattering channels, in which the neutron is re-emitted with reduced energy and the target nucleus is raised to an excited state. These channels depend upon the existence of suitable levels in the target nucleus.

(c) Fundamentally similar to the elastic channel, but inhibited by the potential barrier, is the proton-emission channel. This leads to (n, p) reactions, which can be produced by fast neutrons in almost any target element. The residual nucleus, however, contains a neutron where the target nucleus contained a proton; thus it is often unstable and liable to beta decay.

(d) If the incident energy is high enough, two neutrons may be emitted; (n, 2n) reactions of this type may leave residual nuclei which are either stable or positron-active through deficiency in neutrons.

(e) Another particle emitted fairly frequently is the alpha-particle; (n, α) reactions sometimes represent a favourable redistribution of energy, especially in the lighter elements.

(f) If all the particle-emission channels fail, the energy of excitation can be emitted as a gamma ray; but (n, γ) reactions are less important for fast neutrons than for slow.

Actual reactions. In light elements, fast neutrons may cause complete disintegration, or (n, α) reactions. An example in which these amount to the same thing is the disintegration of carbon into three alpha particles,

$$^{12}_{6}\text{C} \longrightarrow 3\,^{4}_{2}\text{He} \quad (Q = -7 \cdot 34 \text{ MeV}).$$

In medium-sized nuclei, the (n, p) and (n, 2n) reactions become important, especially since many of them give unstable products with recognizable half-lives which can be used in the measurement of fast neutron fluxes. Each can occur only above a threshold energy, which is the kinetic energy required to make the reactions occur; for a reaction with negative Q, the threshold energy is the positive energy

$$-\frac{M+m}{M}Q,$$

where M is the target mass and m the neutron mass.
Examples are

$$^{56}_{26}\text{Fe} + \,^{1}_{0}\text{n} \longrightarrow \,^{1}_{1}\text{H} + \,^{56}_{25}\text{Mn} \quad \text{(threshold } 2 \cdot 1 \text{ MeV)}$$
$$\Big\downarrow 2 \cdot 6 \text{ hours}$$
$$\beta^{-} + \,^{56}_{26}\text{Fe} \quad (E_{\max} = 2 \cdot 86 \text{ MeV)};$$

$$^{107}_{47}\text{Ag} + \,^{1}_{0}\text{n} \longrightarrow 2\,^{1}_{0}\text{n} + \,^{106}_{47}\text{Ag} \quad \text{(threshold } 9 \cdot 6 \text{ MeV)}$$
$$\Big\downarrow 2 \cdot 45 \text{ min}$$
$$\beta^{+} + \,^{106}_{46}\text{Pd} \,(E_{\max} = 1 \cdot 95 \text{ MeV)}.$$

Finally, we must point out that in the largest nuclei fast neutrons can cause fission. This is not the most important mechanism for bringing about fission in the ordinary type of reactor (see Chapter 9) which depends upon slow neutrons, but in the more advanced 'fast' reactors an important contribution is made by the fast neutrons.

8.4.2 *Reactions of slow neutrons*

Particle-emission reactions. When a slow neutron is absorbed by a target nucleus to form a compound nucleus the excitation, though high, is less extreme than when kinetic energy also is provided by the incident neutron. The subsequent de-excitation of the compound nucleus is therefore correspondingly slower, and the properties of the individual nuclear species have more opportunity to affect the outcome of the reaction.

In many elements there is a reaction channel with positive energy release Q which allows some of the energy of excitation to appear as kinetic energy of the reaction products. Examples in the light elements are

$$^6_3\text{Li}(n, \alpha)^3_1\text{H} \quad (Q = 4\cdot785 \text{ MeV}),$$

$$^{10}_5\text{B}(n, \alpha)^7_3\text{Li} \quad (Q = 2\cdot791 \text{ MeV}),$$

$$^{14}_7\text{N}(n, p)^{14}_6\text{C} \quad (Q = 0\cdot626 \text{ MeV}).$$

The first two of these are used for the detection of slow neutrons (see section 8.5). The third, the (n, p) reaction in nitrogen, is of biological importance: through it, slow neutrons in nitrogenous material can produce heavy ionization along short tracks, which can cause biological damage if they cross the sensitive part of a living cell. Apart from this reaction, slow neutrons do not have any specific biological effects; their other effects are through capture reactions leading to gamma emission, which in turn leads to weak ionization along the tracks of the electrons from Compton scattering of the gamma rays. The biological effects of slow neutrons are thus similar to those of gamma rays and slow protons produced internally throughout the irradiated material.

Radiative capture. There are many elements in which de-excitation following slow neutron capture can occur only by gamma emission. An example is the radiative capture of slow neutrons by hydrogen

$$^1_1\text{H}(n, \gamma)^2_1\text{D} \quad (Q = -2\cdot225 \text{ MeV})$$

This process has already been mentioned as limiting the usefulness of ordinary water as a moderator (see section 8.3.1, under *Moderating ratio*). It is of course the inverse of photodisintegration of the deuteron (section 4.3.3).

In some of the larger nuclei (n, γ) reactions have positive rather than negative practical value. For example, indium-115 is converted to indium-116, which has a characteristic half-life of 54·1 minutes for the subsequent beta and gamma decay. This makes activation of indium a convenient tool for measuring slow neutron fluxes.

Another important (n, γ) reaction is

$$^{113}\text{Cd}(n, \gamma)^{114}\text{Cd}.$$

This is important because it has so large a cross-section (over a thousand barns) for all neutron energies below 0·2 eV, that a sheet of cadmium metal is an effective screen against slow neutrons.

In many medium-sized nuclei (n, γ) reactions occur through formation of definite levels in the compound nucleus. When this happens the cross-section shows a resonance at the appropriate neutron energy. A few such resonances occur in the thermal or epithermal range of energies, but most are in the so-called resonance range, from 1 keV to 1 eV (see section 8.1.2).

A final, important reaction induced by slow neutrons is the fission of heavy nuclei. This reaction is the basis of Chapter 9.

8.5 Neutron detectors

Neutrons, being uncharged, cannot be detected directly. Instead they must be allowed to undergo some nuclear interaction, either scattering or a reaction. If some of the products are charged and are moving fast enough to cause ionization, these can be detected, and the presence of the neutrons inferred.

8.5.1 *Fast neutron detectors*

Proton recoil. Most instruments for detecting fast neutrons depend on elastic scattering by protons in hydrogenous material, the recoil protons being detected.

If the direction of the incident neutrons is known, and that of the recoil protons measured or limited, this process may be used for measuring the energy distribution of the neutrons, since the kinetic energy of a proton recoiling at an angle ϕ is a fraction $\cos^2\phi$ of that of the incident neutron (see section 3.7.1). Either ϕ may be measured for individual protons or the proton detector may be arranged to collect protons only over a specified range of values of ϕ. The former method allows calculation of the energy of individual neutrons, while the latter introduces a spread in the calculated neutron energies of width given by the uncertainty in $\cos^2\phi$.

Alternatively, the detector may be insensitive to the direction of the recoil proton, so that it gives the distribution in energy of all the recoil protons. For a single group of neutrons, this distribution spreads from zero up to the neutron energy. The general features of a neutron spectrum can be inferred from a continuous distribution of this type, but the finer details tend to be obscured since forward recoils from less energetic neutrons give protons of the same energy as wide-angle recoils from the more energetic neutrons.

In cases where neutron flux is isotropic, the direction of the recoil proton contains no useful information, and all detectors give continuous distributions of the type mentioned above.

The most difficult case is that of a neutron flux which is neither isotropic nor unidirectional. To obtain information on the energy spectrum of such a flux, as a function of direction, one must use a detector which gives values for the direction and energy of the recoil protons, and use statistical methods to unfold the distributions of neutron energy and direction. Without the information on direction of the recoil protons, their energy distribution serves only to fix limits for the neutron spectrum.

Instruments using recoil protons. Various geometrical configurations are possible for fast neutron detectors depending on recoil protons. These include the following:

(a) A target of material such as polyethylene is placed in front of an instrument for detecting the recoil proton, for example, an ionization chamber or a scintillation counter. If the detector is thick enough to stop the recoil protons, it can yield a distribution of the energies with which they leave the target; for a thick target this is not the same as the distribution of recoil energies, but the latter can be inferred, and hence within limits of accuracy set by the angle of acceptance of the detector the neutron spectrum deduced.

(b) The proton detector itself may contain hydrogen, and act as target as well as detector. If the detector is a plastic scintillator attached to a photomultiplier tube, or a gas counter containing hydrogen, it will have little or no power to distinguish between different directions of proton recoil. On the other hand, the distribution of pulse heights will give a true representation of the distribution of proton energies, without any spread from target thickness (though some spread can be given by escape of protons from the surfaces if the detector is not much larger than the range of the fastest protons). The distribution of neutron energies can be deduced from the distribution of pulse heights as discussed in the previous section *Proton recoil.*

(c) A visual detector containing hydrogen can act as target, while providing information of the direction and the energy of individual recoil protons. Early work on neutron spectra from nuclear reactions was done with cloud chambers. With a small neutron source outside the chamber, the direction and length of the track of each recoil proton could be used to give a value for the energy of an individual neutron, the relation between range and energy for protons in the particular gas having been determined by other experiments.

Following this early work in cloud chambers, photographic plates have proved useful as neutron detectors. The hydrogen is contained in the gelatin of the photographic emulsion, and the range and direction of individual recoil protons in the emulsion give the energies of the individual scattered neutrons in experiments with good geometry. For monitoring the general intensity of fast neutrons to which personnel are exposed, the number and range of recoil proton tracks in a photographic emulsion provides a convenient quantitative criterion. This is widely used to supplement the information on general gamma irradiation obtained from blackening of an ordinary photographic film.

Threshold detectors. Activation reactions tend to have rather low cross-sections for fast neutrons, but there are occasions on which a single measurement of the activity produced in a chosen target material can be a convenient way of obtaining a value for the flux of neutrons with energy above the threshold for the particular reaction. The reaction is identified by the characteristic half-life of the product for β^--, β^+- or γ-decay. Examples of two reactions which can be used in this way have been mentioned in section 8.4.1, under *Actual reactions.* Reactions of the (n, 2n) type, for example in ^{12}C, ^{14}N, ^{19}F, ^{50}Cr, ^{58}Ni, ^{107}Ag and ^{127}I, give a range of thresholds from 9 to 20 MeV, with half-lives from 10 minutes to 16 days, while (n, p) reactions in ^{24}Mg, ^{27}Al, ^{31}P, ^{32}S, ^{52}Cr and ^{56}Fe give lower threshold energies, from 1·0 to 2·8 MeV, with half-lives from 3·9 minutes to 14 days.

(n, α) *reactions.* For the detection of slow neutrons, an exothermic reaction is needed, to provide charged particles moving fast enough to cause ionization. For instantaneous recording of slow neutron flux, the two most commonly used reactions are $^{10}_{5}B(n, \alpha)$ $^{7}_{3}Li$ and $^{6}_{3}Li (n, \alpha)$ $^{3}_{1}H$. In both of these reactions, the two product particles are emitted in opposite directions and produce strong ionization which can give large pulses in a scintillation counter or a proportional counter. Proportional counters and ionization chambers containing the gas boron trifluoride, and scintillators loaded with boron or lithium compounds, are all used in commercially available slow-neutron detectors.

The reaction in boron-10 is particularly satisfactory, since its cross-section is proportional to $1/v$, over a wide range of values of the neutron velocity v. This means that when there is a whole spectrum of thermal neutrons, each group causes a number of counts per second proportional to

$$N(v) \, \sigma(v) v = \text{constant} \times N(v),$$

where $N(v)$ is the number per unit volume with velocity v. Thus the total counting-rate is proportional to the total number of neutrons per unit volume.

Activation methods. When instantaneous recording is not required, an activation reaction such as $^{115}In(n, \gamma)^{116}In$ (see section 8.4.2, under *Radiative capture*) may be used. Different (n, γ) reactions have cross-sections varying differently with neutron energy, so a choice appropriate to the particular need must be made. Especially useful are reactions having a strong resonance at a particular neutron energy, since they give activities which are virtually measures of the neutron flux at the resonance energy.

Further reading

L. F. Curtiss, *Introduction to Neutron Physics*, 1959, Van Nostrand.
S. Glasstone and M. C. Edlund, *Elements of Nuclear Reactor Theory*, 2nd edn, 1960, Van Nostrand.
D. J. Hughes, *Neutron Optics*, 1954, Interscience.

Chapter 9
Nuclear Reactors

9.1 Nuclear stability

9.1.1 *General considerations*

The word 'stability' is used in two distinct but partly overlapping senses. First it may be used in the sense of binding energy or negative stored energy, to compare the energy content of two forms of matter. Thus when we say that the system consisting of two alpha particles is more stable than the system in which the same nucleons are arranged as a free proton and a $^7_3$Li nucleus, we are saying that the total binding energy is greater for the two alpha particles, but we are saying nothing about the conditions under which the one system will transform itself into the other. The second sense refers to the question of whether or not spontaneous transitions occur. In this sense a $^7_3$Li nucleus is stable, as is an alpha particle, because they have no spontaneous decay processes, a $^{238}_{92}$U nucleus is unstable because it emits an alpha particle in a decay process with a half-life 0.14×10^{18} s, and a free neutron is even more unstable because it undergoes beta decay with a half-life of 650 s.

Of course the second type of instability presupposes some measure of the first type, but the first does not imply the second unless a mechanism is available for the transition which might release the stored energy. In the rest of this section, we shall discuss the first type of stability. In later sections we shall consider mechanisms for transitions that may result from differences of binding energy.

The table of masses given in the Appendix may be used to give a general picture of the relative stabilities of different types of nucleus, by calculating from the mass of each nucleus the total binding energy with respect to the appropriate number of free neutrons and protons. Then, dividing by the mass number A, we get the binding energy per nucleon, which is a measure of the over-all stability of the nucleus (stability in the first sense). Values of the binding energy per nucleon, calculated for the stablest isotope of each element, are plotted against A in Figure 34(a). This figure shows clearly that the stablest configuration of nuclear matter (in the sense of the most tightly bound) is that found in the medium-sized nuclei. Figure 34(b) shows, for comparison, the mass excess $M - A$, which measures the total stored energy with respect to nucleons in a standard nucleus (^{12}C).

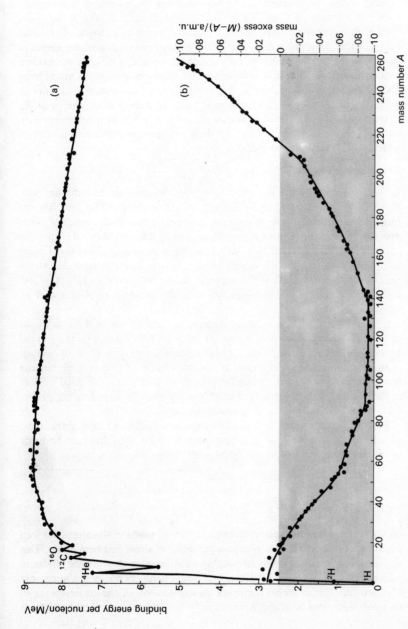

Figure 34 Nuclear stability plotted for most abundant (or longest-lived) isotope of each element, expressed in terms of (a) binding energy per nucleon, and (b) mass excess

9.1.2 Spontaneous decay

For the present discussion, the interesting point in Figure 34 is the decrease of binding energy per nucleon as we approach the largest nuclei. This results from the increase of Coulomb energy with increasing Z, and it manifests itself in an increasing proneness to spontaneous decay by emission of an alpha particle. This does not release more than a small proportion of the excess of stored energy, but for most naturally occurring nuclei it is the only mechanism available for transition to a state of smaller Coulomb energy. In many large nuclei, even this mechanism is not available, and beta decay may occur, balancing an increase in Coulomb energy against a decrease in energy of the strong interaction. In fact all known nuclei with $Z > 84$ decay spontaneously, either by alpha emission, or by beta decay or electron capture to a product which itself emits an alpha particle.

9.1.3 Transuranic elements

Beyond the limit of the naturally occurring elements, set by uranium with $Z = 92$, artificial transuranic elements have been made. The first few are formed by neutron capture in uranium, followed by beta decay, which converts one of the neutrons to a proton, thereby increasing Z by one. Beyond plutonium ($Z = 94$), still larger nuclei have been made by bombarding plutonium targets with alpha particles, carbon nuclei, or even ions as heavy as neon-22. At the time of writing, the limit is the element hahnium, with $Z = 105$, $A = 260$, made by bombarding californium-249 with nitrogen-15 ions (Ghiorso *et al*, 1970).

It is possible that special circumstances may make some still-larger nuclei capable of existing without spontaneous decay for long enough to be observed as new elements in the future. These circumstances will have to include especially favourable arrangement of the extra nucleons to give large binding energy in the strong interactions between them, and also some good fortune in the relative binding energies of neighbouring isotopes, to reduce the probability of transitions leading to spontaneous decay.

However, it is certain that the attempts to manufacture new transuranic elements will continue, for the experimental limits of techniques both for producing them and for observing them are continually advancing.

9.2 Fission

9.2.1 The discovery of fission

In some of the largest types of nucleus, a further mode of disintegration can occur, in addition to the long-known processes of alpha and beta decay. This is fission into two fragments of roughly equal mass, a process the effects of which were first observed by Hahn and Strassmann (1939). In the same year Meitner and Frisch (1939) suggested an explanation of the observations in terms of fission, and Bohr and Wheeler (1939) published a theoretical description based on a liquid-drop model of the nucleus as follows. (See also the review by Fraser and Milton, 1966.)

9.2.2 *Conditions for fission*

In the analogy between a nucleus and a drop of liquid, it is assumed that forces between particles in the interior are saturated, while unsaturated forces on the surface give rise to a surface tension; that is, to a surface energy proportional to the area of the surface. In a liquid drop, the need to minimize this surface energy gives the drop its spherical shape, which is maintained until the drop is distorted by external forces or by internal oscillation. In a nucleus of given volume, the energy contains two terms which are dependent on the shape, the first corresponding to the surface energy, and the second to the Coulomb energy of the charge distribution.

The surface energy increases with distortion of the nucleus from its spherical shape, while the Coulomb energy decreases, since the average separation of the protons is increased. The calculation of Bohr and Wheeler showed that, for nuclei with Z^2/A less than a certain limit, the total energy would be increased by distortion. This is the situation in a normal nucleus, which is stable against interactions that might tend to distort it from its normal spherical shape. A nucleus with Z^2/A greater than the limit would have its total energy decreased by distortion, so if it were ever created it would disintegrate on the smallest perturbation of its spherical shape. In fact the known nuclei have Z^2/A rising from a value of one for hydrogen and helium to thirty-six for uranium-235. From the observation of fission in uranium-239, formed by neutron capture in uranium-238, Bohr and Wheeler estimated that the limit of stability with respect to distortion was at a value

$$\frac{Z^2}{A} = 47 \cdot 8.$$

$\qquad$ **9.1**

Extension of the argument shows that, if we consider distortion of a large nucleus carried to the point of creating two separate half-sized nuclei, the total energy will then have decreased, because medium-sized nuclei have greater binding energy per nucleon than the very large ones. Thus, if the distortion is imagined to occur through separation of one spherical charge distribution into two (at first overlapping) nearly spherical charge distributions, the energy must vary with the separation as shown in Figure 35. At first the energy rises, because Z^2/A, though large, is less than 47·8, but after reaching a peak it must drop off towards the value representing the energy of two independent smaller nuclei. If the peak is not too wide, some quantum-mechanical tunnelling can occur, giving spontaneous fission analogous to ordinary alpha decay through quantum-mechanical penetration of a potential barrier. Among the naturally occurring nuclei, spontaneous fission has been observed in $^{238}_{92}$U, $^{235}_{92}$U and $^{232}_{90}$Th, but the probability is low, for all have mean lives greater than 10^{16} seconds, against the combined effects of spontaneous fission and other modes of decay.

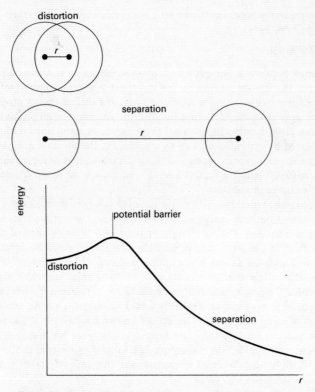

Figure 35 Energy of a distorted nucleus and of departing fission fragments, showing potential barrier for fission

Spontaneous fission with shorter mean life is now known to occur in many of the man-made transuranic elements – for example, $^{239}_{94}$Pu, $^{241}_{95}$Am, $^{252}_{98}$Cf, $^{256}_{100}$Fm and $^{256}_{102}$No.

Fission occurs even faster in some of the short-lived, highly excited nuclei which are formed by neutron capture. But when the fission occurs in a nucleus that lives only for a time of order 10^{-20} second, it is no longer called spontaneous fission, and is regarded as a reaction channel characteristic of the interacting initial particles. It is then treated as neutron-induced fission of the target nucleus. The example, uranium-239, formed in an excited state by capture of a fast neutron in uranium-238, has already been mentioned. Especially important are the three relatively long-lived nuclides in which slow-neutron capture can lead to a fission reaction; these are uranium-235, plutonium-239 and uranium-233 (an isotope of uranium unknown in nature, but prepared by breeding from thorium-232, see section 9.6).

9.2.3 The emission of other particles in fission

The liquid-drop model has a further qualitative success, in that a vibrating liquid drop forms a neck of liquid as it separates into two, and occasionally a part of this neck becomes detached and flies off as a droplet independent of the two main fragments. This provides an analogy, if not a detailed mechanism, for the alpha-particle emission which occurs in about 0·2 per cent of fission processes. In fact a whole range of small fragments is found as an occasional third particle emitted in fission.

At the lower end of this range are the free neutrons which are inevitably emitted in the fission process itself, and by subsequent rapid decay of unstable fission fragments. The list of nuclei shows that the proportion of neutrons in a fissionable nucleus is higher than in stable nuclei of half its size. Thus even if the fission fragments are the heaviest isotopes of the elements corresponding to their values of Z, there is still an excess of neutrons.

Of these excess neutrons many are emitted as 'prompt' neutrons, within about 10^{-14} s of the fission process, but some remain in neutron-rich fission fragments which are converted by beta decay to nuclei with a smaller neutron: proton ratio.

9.2.4 Delayed neutrons

It is observed that some neutrons are delayed, appearing seconds or even minutes after the fission process. This cannot result from direct decay of a neutron-rich fission fragment, for neutron emission faces no potential barrier and must occur in a time of order 10^{-20} s if it is to occur at all.

The explanation of the delayed neutrons is found in beta decay of a fission fragment, with characteristic mean life, to another nucleus which is still so rich in neutrons that, in the highly excited state in which it is sometimes formed, it is unstable with respect to neutron emission.

A neutron is then emitted immediately the intermediate nucleus is produced (i.e. within about 10^{-20} s), but the over-all time scale is controlled by the preceding beta decay. Groups of delayed neutrons are in fact observed with different delay times characteristic of individual beta-decay processes.

The practical importance of delayed neutrons, in allowing control of reactors, is discussed in section 9.4.2.

9.2.5 The energy released in fission

The plot of binding energy per nucleon in Figure 34 shows that when the nucleons of a nucleus with $A \simeq 240$ are rearranged as two nuclei each with $A \simeq 120$, energy of order 1 MeV per nucleon should be released.

In fact the emission of free neutrons reduces the energy release by about 8 MeV per free neutron emitted, and a characteristic value for the total energy released in fission is 200 MeV, with about 165 MeV appearing as kinetic energy of the fission fragments, and 35 MeV as kinetic energy of neutrons and energy of the gamma rays, electrons and neutrinos from decay of the fission fragments.

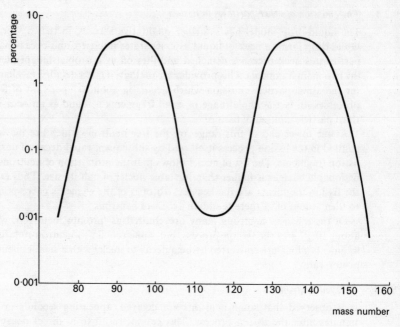

Figure 36 Distribution of mass of fission fragments from uranium-235

9.2.6 *Distribution of masses of fission fragments*

It is a curious fact, not yet fully explained, that fission tends to occur asymmetrically, into two fragments of unequal size. The commonest mode of fission is into a heavy fragment with $A \simeq 140$ and a lighter one with $A \simeq 95$. An actual distribution of masses of fission fragments is shown in Figure 36, in which the depth of the minimum between the two peaks shows that the bias against symmetrical fission amounts to a factor of order two hundred.

9.3 Multiplication factor

To obtain energy in useful quantities from nuclear reactions, it is necessary to set up a system in which there is a mechanism for the individual nuclear reactions to bring about further reactions. For the process to be self-sustaining, the average number of reactions brought about by each preceding reaction must be at least one.

These conditions can be realized in the fission process, in which the emitted neutrons can provide the mechanism for causing further nuclei to undergo fission reactions. In the thermal reactor, use is made of the high fission cross-section of uranium-235 or plutonium-239 for slow neutrons. The fast neutrons emitted in each generation of fission processes are slowed down to thermal energies in a moderator, and are then able to interact with more nuclei to cause the next generation of fission processes. In such a reactor, the multiplication factor k is defined as the average number of slow neutrons in one generation per slow neutron absorbed by fissile material in the preceding generation. This definition is equivalent to the alternative one in terms of numbers of reactions.

The average number of fast neutrons emitted each time a slow neutron capture leads to fission may be given the symbol v, for which a typical value is just over two. The average number of fast neutrons η emitted per neutron absorbed in a fissile nucleus, is smaller than v by a fraction given by the cross-sections for absorption with or without fission:

$$\eta = v\frac{\sigma_{\text{fission}}}{\sigma_{\text{total}}}. \qquad \qquad 9.2$$

A few of these fast neutrons may collide with fissile nuclei, to bring about further fast-neutron-induced fission. This has the effect of increasing slightly the total number of fast neutrons. The increase may be described by a fast-fission factor ε, with typical value about $1 \cdot 01$. There are now $\eta\varepsilon$ fast neutrons, and of these a fraction p will escape capture in the process of being moderated to thermal energies. Finally, a fraction f of the thermal neutrons will be absorbed by fissile nuclei, the others being absorbed in the moderator or other material. If these are the only considerations, the multiplication factor is

$$k = \eta\varepsilon pf. \qquad \qquad 9.3$$

There is in fact another important consideration: if the system is not infinitely large, some neutrons will be lost through its surfaces. We may allow for this by including a factor $1 - l$, where l is the fraction lost, or we may treat the loss of fast neutrons and that of slow neutrons separately. If a fraction l_{f} of the fast neutrons is lost, and a fraction l_{s} of the slow neutrons, the over-all multiplication factor is

$$k = \eta\varepsilon pf(1 - l_{\text{f}})(1 - l_{\text{s}}). \qquad \qquad 9.4$$

Since the condition for a self-sustaining reaction (often called a chain reaction) is

$$k \geqslant 1, \qquad \qquad 9.5$$

its realization depends on obtaining high enough values for all these factors, which we shall now discuss separately.

For fission of a given type of nucleus, η is fixed. Some actual values are given in Table 11.

The fraction moderated p depends on having a moderator with a high logarithmic decrement ξ, and small capture cross-section. So far as the moderator is concerned, p is effectively given by

$$p = 1 - \frac{18\cdot 2}{\text{moderating ratio}} \qquad \qquad 9.6$$

Table 11 Numbers of Neutrons per Fission

		^{235}U	^{239}Pu	^{233}U	^{238}U	^{232}Th
Thermal *neutrons*	v	2·47	2·90	2·51	–	–
	$\eta - 1$	1·07	1·10	1·28	–	–
Fast *neutrons*	v	2·52	2·98	2·59	2·61	2·34
	$\eta - 1$	1·18	1·74	1·42	–	–

v is the average number of neutrons emitted per fission;

η is the average number of neutrons emitted per neutron absorbed;

$\eta - 1$ is the average number of neutrons per absorption in excess of those needed to maintain the chain reaction, and equals the upper limit to the breeding ratio.

(see section 8.3.1 under *Moderating ratio*). Capture in structural materials or impurities (including any accumulated fission products, will reduce the over-all value of p. It follows that a high value of p requires a high moderating ratio, and also that there should be not too many impurities or other materials present which are capable of capturing neutrons. Sometimes p is known as the 'resonance escape probability,' a name which points out a further important condition for obtaining a high value of p. If the fissile nuclei, or any other nuclei with them, have a resonance for non-fission capture of neutrons at an intermediate energy, something must be done to prevent interactions at this energy. The usual device for doing this is the heterogeneous reactor, which contains fissile material (fuel) in the form of rods regularly spaced in a moderating matrix. The spacing between the rods is chosen to give the neutrons a high probability of being slowed down to thermal energy in the moderator before they enter another fuel rod. Thus the neutrons are in the moderator when they pass through the energy at which absorption might have occurred, and when they re-enter the fissile material, fission has to compete only with the non-fission capture cross-section at the same thermal energy (a competition covered by the ratio η/v).

The fast-fission factor ε is a constant not much greater than one. In a heterogeneous reactor, it depends on the size of the fuel rods, more fission being induced by fast neutrons if the rods are large.

Unlike p and ε, f tends to be reduced by concentrating the fuel into rods in a heterogeneous reactor. This is because the density of slow neutrons tends to be higher in the moderator than in the rods, a situation which favours unproductive capture in the moderator.

Thus the choice of a configuration to maximize εpf involves compromise between εp on the one hand and f on the other.

The fractions lost, l_f and l_s, may be considered as final parameters, which may be adjusted by choosing the size of the reactor, after the choice of materials and geometrical configurations have fixed the value of $\eta\varepsilon pf$ at some value greater than one. For a given value of $\eta\varepsilon pf$, the size of the system can be chosen to give k any required value less than that of $\eta\varepsilon pf$. The size which makes $k = 1$ is called the critical size, because a chain reaction will spread in a larger system, but die out or fail to start in a smaller.

9.4 Control of reactors

9.4.1 *Method of control*

The basic technique for controlling a nuclear reactor is insertion and withdrawal of rods of material like cadmium, which absorbs slow neutrons and thus can change the average value of p for the reactor as a whole, hence varying k. If the reactor has a value of k greater than one with the control rods withdrawn, the reaction rate can be allowed to build up to a desired level, and then the control rods pushed in far enough to make $k = 1$. Switching off can be done by pushing them farther in, to make $k < 1$.

The time scale of the response to a movement of the control rods is important, as is the rate of increase of reaction rate when $k > 1$. The average time taken by a neutron in a thermal reactor, from its moment of emission to the moment at which it stimulates a new fission is of order one millisecond, which is mostly taken up on the later stages of moderation and in diffusion at thermal velocities. At a characteristic thermal velocity of $2200\ \mathrm{ms^{-1}}$, this time represents diffusion for a total (travelled) distance of 2·2 m.

Suppose $k = 1·01$, so that the neutron flux rises by 1 per cent per generation, that is, 1 per cent per millisecond. Over many generations, the neutron flux and the rate of reaction will rise exponentially with time, the time constant in this example being 0·1 s. A reaction rate increasing exponentially with this time constant could not be controlled safely with mechanically operated cadmium rods. Even with $k = 1·001$, and a time constant of one second, a fast-working servo-mechanism would be needed to move the control rods.

9.4.2 *The importance of delayed neutrons*

In fact the control of a thermal reactor is not so difficult as the preceding paragraph suggests, because not all the neutrons follow the time scale indicated. The latter is characteristic of the prompt neutrons, but we must consider also the delayed neutrons (see section 9.2.4). If the multiplication factor k has a value less than one for the prompt neutrons alone, but greater than one when the delayed neutrons are included, any change of reaction rate or response to movement of a control rod will occur on a time scale set by the delayed neutrons, as illustrated in Figure 37. These are emitted in groups with characteristic mean delays ranging up to over a minute, which in near-critical reactors lead to time constants of many hours.

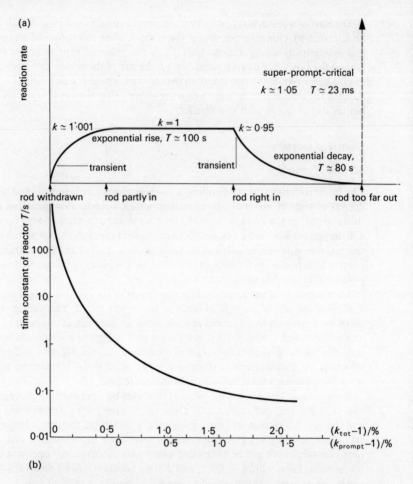

Figure 37 Time scale for control of thermal reactor using uranium-235.
(a) Response to movement of a control rod, showing transient followed by exponential change of reaction rate. (b) Time constant T for increase of reaction rate as a function of $k-1$

9.4.3 *Temperature coefficient*

An essential condition for the safety of all reactors is that the multiplication factor k should decrease as temperature rises. Most designs achieve the necessary negative temperature coefficient of k through such factors as thermal expansion of the fuel, but it has been found advisable to exclude certain forms of construction in order to avoid the possibility of achieving a positive temperature coefficient, which could lead to a dangerously divergent reaction.

9.5 Reactor using natural uranium

Natural uranium contains two principal isotopes, of which uranium-238 is the more abundant (about 99·3 per cent), while uranium-235 is present to the extent of only 0·7 per cent. Uranium-235 has a high fission cross-section (590 barns) for thermal neutrons, with a radiative capture cross-section of 108 barns. Uranium-238, on the other hand, has a threshold at 1·4 MeV for neutron-induced fission, and has strong resonances for neutron capture at intermediate energies, from 1 keV to 5 eV, which give uranium-239 in a state which decays to plutonium-239 by two successive beta decays via $^{239}_{93}$Np. Thermal neutrons do not interact strongly with uranium-238.

The naturally occurring mixture of uranium-235 and uranium-238 thus has precisely the properties for which a heterogeneous reactor is needed. Fission of uranium-235 can be induced in rods of metallic uranium by slow neutrons.

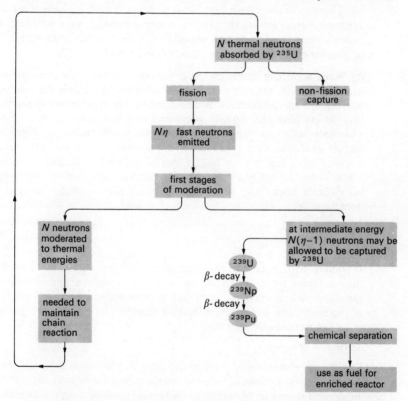

Figure 38 Principle of the breeder reactor. Table 11 gives $\eta = 1·07$. Ideally, $\eta - 1$ nuclei of plutonium-239 are obtained as by-products in the fission of each nucleus of uranium-235

The emitted fast neutrons go out into the moderator, where they are slowed down to an energy below the lowest resonance for capture in uranium-238. Then as thermal neutrons they can diffuse into the uranium rods again, to cause another generation of fission processes in uranium-235. The commonest moderator for this type of reactor is carbon. Heavy water (D_2O) can be used, but ordinary water causes too many neutrons to be lost by the capture reaction $^1H(n, \gamma)^2H$ (see section 8.4.2, under *Radiative capture*).

In the cycle as described above, uranium-235 is the only fuel, and the uranium-238 is a spectator, rendered inert by the action of the moderator in slowing down the neutrons beyond the energy at which they might have been captured by uranium-238. In first approximation, the natural-uranium reactor is indeed a device for obtaining a chain reaction in uranium-235 in the presence of uranium-238. There are two developments from this which can be useful:

(a) A fraction of order 1 per cent of fast neutrons can cause fission of uranium-238 nuclei before leaving the uranium rod in which they were produced. This causes a little uranium-238 to be used up, giving a little extra energy, and helping the chain reaction by giving a value of ε of about 1·02.

(b) When other conditions are good enough to allow some neutrons to be used in this way, the geometry may be adjusted to increase the capture of intermediate-energy neutrons by uranium-238. The product of two successive beta decays after this capture is plutonium-239, which, like uranium-235, undergoes fission on capture of a slow neutron. A reactor in which such capture processes have been deliberately increased is known as a breeder reactor, since it breeds useful fuel (^{239}Pu) from the otherwise useless uranium-238, as illustrated in Figure 38. Since plutonium is a new element, with properties different from those of all other materials present, it can be separated from the uranium afterwards by chemical means.

9.6 Enriched reactors

9.6.1 *Advantages*

If uranium with more than the usual proportion of uranium-235 is available, the conditions for a chain reaction are easier to achieve, and consequently there is more freedom of choice in such matters as over-all size and properties of the moderator. For example, smaller reactors can be built and ordinary water can be used as moderator.

9.6.2 *Separation of isotopes*

Enrichment of uranium in uranium-235 requires a laborious sequence of small enrichments by gaseous diffusion in uranium hexafluoride, for the difference in physical properties of uranium hexafluoride resulting from the difference in mass of the two uranium isotopes is extremely small. However, it can be achieved at the cost of correspondingly large quantities of space and power.

As alternatives to gaseous diffusion, thermal diffusion and, more recently, ultra-centrifuging have also been used.

9.6.3 *Breeding*

For most purposes plutonium may be used instead of uranium-235. It may be obtained by breeding from uranium-238 in a reactor, using either natural uranium or already enriched fuel. If the reactor is operated for the sole purpose of producing slow-neutron-fissile material, the cost is not very different from that of separating the uranium isotopes, but this situation is now past, and the plutonium is usually a by-product reducing the cost of the power which the reactor was built to generate.

With uranium-235 and plutonium-239 as possible fuels for slow-neutron reactors, and with uranium-238 in natural uranium available as a starting point for breeding plutonium-239, breeder reactors can be made to produce more fuel as each batch is used. Thus by passing plutonium from reactor to reactor, the energy stored in the uranium-238 of natural uranium can be released in addition to that of the rarer uranium-235.

The amount of breeding is limited by the number of neutrons available in excess of those needed to maintain the chain reaction. With an average of v neutrons emitted per fission, and $\eta = v\sigma_{\text{fission}}/\sigma_{\text{total}}$ neutrons emitted per neutron absorbed, one out of η being needed to maintain the chain, the greatest possible number of neutrons available for breeding is

$$\eta - 1.$$

Table 11 shows that for thermal neutrons this number is 1·10 (^{239}Pu) or 1·28 (^{233}U). It is therefore theoretically possible to make a thermal reactor breed more fuel than it burns. However, the loss of neutrons in moderation and in capture processes leading neither to fission nor to breeding make the attainable values of breeding ratio (number of fissile nuclei produced per fission) smaller than these ideal values, and so far values exceeding one have been obtained only by breeding uranium-233 from thorium-232.

Thus the best system for thermal breeder reactors, from the point of view of providing fuel for an increasing scale of operation, is fission of uranium-233, with breeding of more uranium-233 by neutron capture in thorium-232.

9.7 Fast reactors

Plutonium differs from uranium-238 not only in undergoing fission on capture of slow neutrons, but also in having a much larger cross-section for fission by fast neutrons. With plutonium available as fuel, it becomes possible to set up a chain reaction with fast neutrons, and to replace the thermal reactor by a fast reactor containing no moderator. This development is important in allowing increased efficiency in the breeding of new fuel from uranium-238, as will be seen from the figures in Table 11. Since η is higher for fast neutrons than for thermal neutrons in plutonium-239, the upper limit for the breeding ratio is $\eta - 1 = 1\cdot74$ for fast neutrons. This is high enough to allow a margin for wastage of neutrons and still give an actual breeding ratio greater than one, much better than can be obtained with plutonium in a thermal breeder reactor.

A typical fast breeder reactor contains a plutonium core, surrounded by natural uranium whose sole purpose is to breed more plutonium by capturing fast neutrons as they escape from the working core. The uranium elements are removed from time to time for chemical extraction of the newly formed plutonium.

So long as the prompt neutrons alone are not enough to make the multiplication factor $k = 1$; that is, so long as the reactor is below *prompt-critical*, delayed neutrons are needed to cause any sustained increase in rate of reaction, and the time scale for the response to control is effectively the same in a fast reactor as in a thermal reactor. Control mechanisms for fast reactors therefore do not need to operate significantly faster, despite the name (which refers to a type of neutron, not to a time constant). The main difference is that absorbers do not have so large an effect in fast reactors, and it is sometimes easier to control the reaction rate by moving a fuel rod or a reflector outside the core, instead of by inserting an absorbing control rod.

9.8   Pulsed reactors

For research purposes, either thermal or fast reactors may be made to give pulses of activity. Normally a reactor has $k \geqslant 1$ only by inclusion of the delayed neutrons, but if the over-all value of k is increased a little, the reactor may become prompt-critical, with $k = 1$ for the prompt neutrons alone, or even *super-prompt-critical*, which has nothing to do with super-promptness, but means above prompt-critical, that is $k \geqslant 1$ for prompt neutrons alone.

Then the rate of reaction will increase with a time constant much shorter than those of reactors depending on delayed neutrons. The mean life of a neutron in a fast reactor, between emission and capture leading to fission is typically 10^{-7} s. Thus if $k_{prompt} \simeq 1{\cdot}01$, the time constant for increase of reaction rate is of order 10^{-5} s (cf. $0{\cdot}1$ s for a prompt-critical thermal reactor, see section 9.4.1).

A built-in safety factor for limiting the increase of rate of reaction in a pulse rising in this way is the negative temperature coefficient of k (see section 9.4.3). This results from several factors, including thermal expansion of the fuel, and Doppler broadening of capture resonances by motion of the target nuclei.

Single pulses may be controlled by the negative temperature coefficient of k, but when repeated pulses are required, they may be obtained by varying k with a rotating wheel which carries an extra piece of fuel in and out of the core of the reactor.

9.9   Heat transfer

As has been implied throughout the preceding discussion, the main reason for setting up nuclear reactors is to use them as sources of energy. The energy released in fission appears first as kinetic energy of the fission fragments and other emitted particles. As the fragments move off into the surrounding matter, they share their energy with it, and the kinetic energy is converted to thermal energy.

An idea of magnitudes is given by the following calculation:

One fission releases 200 MeV $= 3 \cdot 2 \times 10^{-11}$ J.
One kilogramme of natural uranium contains $1 \cdot 8 \times 10^{22}$ ^{235}U nuclei and $2 \cdot 5 \times 10^{24}$ ^{238}U nuclei.
Fission of these ^{235}U nuclei yields energy 6×10^{11} J $= 167$ megawatt-hours.
If the whole of the uranium is used, by breeding, the total yield from one kilogramme is 8×10^{13} J $= 22\,000$ megawatt-hours.

These are considerable quantities of thermal energy, and to carry them away from the fuel rods, even over a long period of time, needs an efficient system of heat transfer. Some fluid has to be circulated through the reactor and a heat exchanger, from which a working fluid can flow to turbines if electrical energy is the objective.

In reactors using natural uranium, the fluid used for heat transfer must have minimum cross-section for neutron capture, and the choices include the gases carbon dioxide and helium, as well as the liquid heavy water (D_2O), which may be used as both moderator and (rather expensive) heat transfer fluid at the same time.

Electrical generating plant installed in Britain up to 1969 has used graphite-moderated reactors, with carbon dioxide under pressure, flowing over uranium metal encased in a magnesium alloy (whence the name 'Magnox reactor'). Plant of 8000 MW capacity, to be installed in Britain from 1970 to 1975, is to be based on 'advanced gas-cooled reactors' working at a higher temperature of about 550°C, made possible by using fuel rods of uranium oxide encased in stainless steel cans. The moderator is still graphite and the heat-transfer fluid carbon dioxide.

Enriched reactors place less stringent requirements on the fluid used for heat transfer. One simple method is to use ordinary water as both moderator and heat-transfer fluid. This method, in which cheapness of construction and cooling are obtained at the cost of using processed fuel, is the basis of the 'boiling water reactor', widely used in American-built generating plant. This system is especially suitable for equipment to supply limited quantities of power in remote places, because an enriched boiling-water reactor can be made small, and transported and installed as a ready-made unit without any heavy engineering on the site.

Fast reactors present different problems of heat transfer, because large quantities of heat are produced in relatively small cores (of order one cubic metre). The favoured solution is to use a liquid metal for heat transfer: mercury, liquid sodium and sodium–potassium mixtures have all been used. It appears very probable that long-term development of nuclear power may depend on fast breeder reactors with liquid sodium for heat transfer.

Further reading

Types of reactor

Review articles in H. R. Hyder (ed.) *Progress in Nuclear Energy*, series 2, vol. 1, 1956, vol. 2, 1961, Pergamon.

Fast reactors

L. J. Koch and H. C. Paxton, *Annual Review of Nuclear Science*, vol. 9, 1959, pp. 473–92.

Economics of nuclear power

J. A. Lane, *Annual Review of Nuclear Science*, vol. 16, 1966, pp. 345–78.

Chapter 10
Thermonuclear Reactions

10.1 Sources of energy

In Figure 34 (p. 135) a plot of binding energy per nucleon for the stable nuclei was used to show that the fission of large nuclei into medium-sized fragments was a potential source of energy.

The same plot suggests, as an alternative source of energy, the fusion of nuclei of hydrogen and the light elements to make larger nuclei. For each proton going from the free state to become part of a nucleus of carbon or something larger, an energy of about 8 MeV would be released. This compares very favourably with the 1 MeV per nucleon released in fission.

But although there are many reactions of protons and deuterons with light nuclei in which the energy release Q is positive, as discussed in Chapter 4, observing them in the laboratory is very different from obtaining useful energy from them. Accelerating machines use large amounts of electrical energy, and any positive energy released in the nuclear reactions which they bring about is liberated as heat in the neighbourhood of the target. The quantity of heat so liberated is usually minute in comparison with the watts of energy dissipated by the beam itself, and even more so in comparison with the kilowatts or megawatts used to run the accelerator and maintain the vacuum in it.

In the fission chain reaction, each fission process emits neutrons which may bring about further fission processes. In this way the reactions may be made to occur in large enough numbers to produce useful amounts of energy. The neutron which provides the link in the chain is a by-product of the reaction, and carries little energy itself; being electrically neutral it can enter a target nucleus without difficulty, thereby bringing it to a state in which fission is almost certain.

In the fusion of light nuclei, a Coulomb barrier must be overcome, and there is no known trick for dodging this. To make fusion reactions occur in large numbers, one must have large numbers of nuclei moving with kinetic energies high enough for reaction to be possible when a collision occurs. The energy released in the reaction must then be fed back into the system to maintain the supply of kinetic energy. If we had stationary target nuclei with incident nuclei moving as a parallel beam, the acceleration of the beam would require far more energy than could be fed back from the energy released by reactions in the target; so since the energy is released as kinetic energy of product particles moving in random directions, we examine the possibility of using the collisions of these particles to raise the temperature of other nuclei, so that random collisions between pairs of these may bring about further reactions.

In other words, we examine the possibility of thermonuclear reactions in gases at high temperatures. The present chapter discusses the principles according to which fusion may occur in thermonuclear reactions, and their application to two situations, namely the stars, where fusion at high temperature is the primary source of energy, and the laboratory, where attempts are being made to achieve controlled fusion at the lowest possible temperature (which is far from low by ordinary standards).

10.2 Principles

10.2.1 *Energy of thermal motion*

When we speak of thermal energies in connexion with the production of slow neutrons, we mean kinetic energies characteristic of thermal motion at room temperature. At a temperature Θ, in kelvins, kinetic energies of thermal motion are of order $k\Theta$, where k is Boltzmann's constant, equal to $8 \cdot 6165 \times 10^{-5}\,\mathrm{eV\,K^{-1}}$. So thermal energy in the above sense means energy of the order $\frac{1}{40}\,\mathrm{eV}$.

We now need to think about thermal motion at much higher temperatures, with $k\Theta$ correspondingly greater. In our discussion of reactions of charged particles, we have so far usually assumed a stationary target nucleus, struck by an incident particle which had previously been accelerated by electrical means to an energy of several megelectronvolts. Artificially produced nuclear reactions have however been observed in beams accelerated to only a few thousand electronvolts. For a collision between two particles of which both are moving, the probability of reaction is given by their kinetic energy of relative motion, that is, the sum of their kinetic energies as measured in their centre-of-mass system. The most probable value of this quantity is in fact equal to $k\Theta$, not $\frac{3}{2}k\Theta$ which is the mean kinetic energy of an individual particle. Therefore we are doing something better than an order-of-magnitude calculation when we say that a convenient unit of temperature for thermonuclear reactions is 10^7 K, which makes

$$k\Theta = 0 \cdot 86\,\mathrm{keV}.$$

In fact the temperatures which we must consider range from 10^6 to 10^9 K. Over the whole of this range, gases of density high enough to be interesting exist in the form of a completely ionized plasma. A plasma is an electrically neutral mixture of independent nuclei and electrons, moving and colliding at random, with no long-lasting association of electrons with nuclei, that is, no atoms, and no shielding of nuclei by atomic electrons.

In Chapter 4, which was devoted to the reactions of light nuclei, it was mentioned that reactions could occur between nuclei of low Z at relatively small incident kinetic energies through quantum-mechanical penetration of the potential barrier, whose height at a given radius is proportional to the product of the charges of the interacting particles.

For a detailed discussion of the quantum-mechanical penetration of potential barriers, the reader is referred to the treatment given by Reid (1972). This treats the problem of emission of alpha particles by radioactive nuclei; the alpha particle can exist with positive kinetic energy inside the potential well which describes the field inside the parent nucleus. Similarly it can exist free outside the nucleus. But between the edge of the potential well and the radius at which the Coulomb potential energy is equal to the total energy available to the alpha particle, it would have to have a negative kinetic energy. Classically, this means that it cannot cross from the inside of the barrier to the outside. But the quantum-mechanical treatment shows that the oscillatory wave function inside the nucleus can be fitted to an exponentially decaying one under the barrier, which in turn fits an outgoing wave beyond the barrier; this is interpreted as a tunnelling through the barrier, resulting in a small probability per unit time that the alpha particle will be found outside, moving away.

For our present purpose, we need the opposite type of barrier penetration. A flux of incident particles approaches a nucleus, and Coulomb repulsion, treated classically, prevents the incident particles from approaching closer than the limit at which the Coulomb energy is equal to the incident kinetic energy. Quantum-mechanical tunnelling can, however, allow a small probability that an incident particle will cross the barrier and enter the region inside the nucleus, where the attractive short-range nuclear forces cancel the electrostatic repulsion and allow the incident particle to exist with positive kinetic energy until it loses its identity by reaction. The probability of barrier-penetration is calculated by fitting incoming spherical waves (representing the incident particle) to a wave function which decreases exponentially as r decreases under the barrier, and then at the boundary of the nucleus fitting this exponentially decreasing wave function to the oscillatory wave function for the region inside the nucleus.

Many of the original calculations in this field were made by Gamow (1938) and by Bethe (1939) and Salpeter (1952). Notations and methods differ, but we shall largely follow those of Thompson (1957).

For barrier penetration in either direction, when E is much less than the barrier height, the principal term in the probability is the Gamow factor

$$\exp\left[\frac{-2\pi Z_1 Z_2 e^2}{\hbar v}\right],$$

in which $Z_1 e$ and $Z_2 e$ are the charges on the two particles, and v is their relative velocity when free of each other's influence. If the particles have masses m_1 and m_2, the barrier-penetration factor takes the form

$$G = \lambda^2 \exp\left[\frac{-2\pi Z_1 Z_2 e^2}{\hbar v}\right],\tag{10.1}$$

where λ is the de Broglie wavelength of a particle of velocity v with mass

$$m = \frac{m_1 m_2}{m_1 + m_2}.\tag{10.2}$$

It was pointed out in section 3.5.3 that m may be called the relative mass, because $E = \frac{1}{2}mv^2$ is the kinetic energy of the relative motion, that is, the sum of the two kinetic energies as measured in the centre-of-mass system. Following equation **3.18** we should point out that to compare figures for a relative energy E with laboratory measurements made with a beam bombarding a stationary target, one must use a bombarding energy

$$E_1 = \frac{m_1 + m_2}{m_2} E.$$

If λ and v are expressed in terms of E, equation **10.1** gives the following expression for the cross-section as a function of E:

$$\sigma(E) = AE^{-1} \exp(-BE^{-\frac{1}{2}}),\tag{10.3}$$

where $\quad B = \pi Z_1 Z_2 e^2 h^{-1}(2m)^{\frac{1}{2}}$

and A is an experimentally determined constant including the partial width for the particular reaction. A has the dimensions of cross-section times energy, while B^2 is an energy. A and B are often called Gamow constants.

Equation **10.3** shows that the cross-section increases rapidly with increasing relative energy E of the reacting particles. It is therefore not possible to work out the reaction rate simply by inserting $k\Theta$ as the most probable value of E. The most probable value of E for all pairs of particles is $k\Theta$, but the most probable value for pairs which react is higher, because of the energy dependence of the cross-section. To resolve this problem it is necessary to consider the actual distribution of velocities.

10.2.3 *The Maxwellian distribution of velocities*

In a gas containing n_1 particles of type 1 per unit volume, each of mass m_1, the velocities are distributed according to the Maxwell law

$$dn_1 = 4\pi n_1 \left[\frac{m_1}{2\pi k\Theta}\right]^{\frac{3}{2}} \exp\left[-\frac{mv_1^2}{2k\Theta}\right] v_1^2 \, dv_1.$$

If there are present also n_2 particles of type 2 per unit volume, each of mass m_2, at the same temperature Θ, their velocities v_2 will follow a similar distribution.

To get the distribution of relative velocities of particles of type 1 with respect to particles of type 2, it is necessary to change the variables, abandoning the individual velocity vectors v_1 and v_2 in favour of the relative velocity v and the velocity of the centre of mass. Integrating over the three components of the velocity of the centre of mass, and over the solid angle in v, we get for the distribution of v (the magnitude of v)

$$n(v)\, dv = n_1\, n_2\, 4\pi \left[\frac{m}{2\pi k\Theta} \right]^{\frac{3}{2}} \exp \left[-\frac{mv^2}{2k\Theta} \right] v^2\, dv, \qquad \textbf{10.4}$$

where m is the relative mass as defined in equation **10.2**.

10.2.4 *The reaction rate*

The number of reactions per unit volume per second is equal to

$$r = n_1\, n_2 \langle \sigma v \rangle, \qquad \textbf{10.5}$$

where n_1 is the number of particles of type 1 per unit volume, and n_2 the number of particles of type 2 per unit volume. $\langle \sigma v \rangle$, which is sometimes called the transition probability, means the average value of σv, taken over all values of the relative velocity v.

If the velocity distribution is given by the Maxwellian formula **10.4**, and the cross-section is given by equation **10.3** with energy dependence resulting from barrier penetration, equation **10.5** for the reaction rate may be rewritten as

$$r = n_1\, n_2 \int\limits_0^\infty \sigma v\, n(v)\, dv$$

$$= n_1\, n_2\, 2\sqrt{\left[\frac{2}{\pi m} \right]} A(k\Theta)^{-\frac{3}{2}} \int\limits_0^\infty \exp - \left[\frac{E}{k\Theta} + BE^{-\frac{1}{2}} \right] dE. \qquad \textbf{10.6}$$

To show which type of collision contributes most to the reaction rate, a cross-section and a velocity distribution are plotted separately in Figure 39. Also plotted is their product, which gives the integrand of equation **10.6**; this has a fairly sharp peak at the energy which makes $(E/k\Theta) + BE^{-\frac{1}{2}}$ a minimum, namely

$$E = E_{\text{opt}} = (\tfrac{1}{2}Bk\Theta)^{\frac{2}{3}}. \qquad \textbf{10.7}$$

The integral in equation **10.6** may be evaluated approximately, with the following numerical result,

$$r = 0.81 \times 10^{-16} n_1\, n_2\, AB^{\frac{1}{3}} M^{-\frac{1}{2}} (k\Theta)^{-\frac{2}{3}} \exp \left[-3 \left(\frac{B^2}{4k\Theta} \right)^{\frac{1}{3}} \right] \text{reactions per cubic}$$

centimetre per second, **10.8**

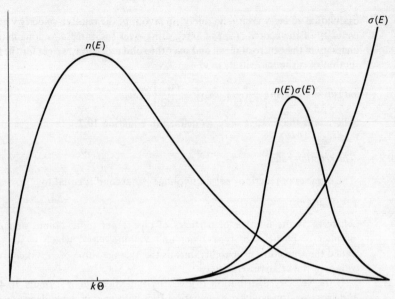

Figure 39 Distributions of energy and reaction rate. E is the kinetic energy of relative motion. $n(E)$ is the number of collisions per unit E for Maxwellian distribution of velocities. $\sigma(E)$ is the cross-section for reaction at energy, dominated by the barrier-penetration factor. $n(E)\sigma(E)$ is the number of reactions per unit E

where n_1 and n_2 are the numbers of each type of particle per cubic centimetre, A is the first Gamow constant, expressed in barns $\times$ keV, B is the second Gamow constant, expressed in $(\text{keV})^{\frac{1}{2}}$, $k\Theta$ is the temperature expressed in keV and M is the relative mass m expressed in atomic mass units.

In equations **10.5–8** $n_1 n_2$ is the number of possible reacting pairs of nuclei per unit volume. In the case of a reaction between two identical nuclei, for example the $\text{d}+\text{d}$ reactions discussed below, the number of possible pairs of nuclei per unit volume is $\frac{1}{2}n^2$; for such a reaction, the equations and arguments of this chapter must be modified by putting $n_1 = n_2 = n$ and inserting the factor one half.

At this point it should be mentioned that although equation **10.8** is basically a theoretical formula, we are usually dependent on experiment for the value of A. The formula may be expected to apply in cases where the motion is genuinely thermal, with a Maxwellian distribution of velocities, and where the cross-section does actually have the ideal energy dependence given by equation **10.3**. But in cases where the experimental determination of A reveals departures from this energy dependence, the measured cross-sections may be put into the calculation of $\langle \sigma v \rangle$, and a better value of the reaction rate calculated. This is especially necessary for the following three reactions, all of which have resonances giving peak cross-section below 1 MeV:

$$^2_1H + ^3_1H \longrightarrow ^1_0n + ^4_2He \quad (Q = 17{\cdot}58 \text{ MeV, peak 60 keV}),$$

$$^2_1H + ^3_2He \longrightarrow ^1_1H + ^4_2He \quad (Q = 18{\cdot}34 \text{ MeV, peak 270 keV}),$$

$$^2_1H + ^6_3Li \longrightarrow 2^4_2He \quad (Q = 22{\cdot}36 \text{ MeV, peak 450 keV}).$$

Here the energies quoted for the peaks are relative energies.

Following the theoretical energy dependence rather roughly, but without any peaks, are the d + d reactions $^2H(d, p)^3H$ and $^2H(d, n)^3He$ (see Chapter 4).

10.2.5   *Mean life in a thermonuclear reaction*

In order to make the discussion of section 10.2.4 quantitative, we refer back to equation **10.5**, which gives r, the number of reactions per unit volume per second, as a product of two factors:

(a) $\langle \sigma v \rangle$, which gives the probability per second of a reactive collision between a specified pair of particles confined in unit volume; it may be measured in $\text{cm}^3 \text{ s}^{-1}$.

(b) $n_1 n_2$ for a mixture of gases, or $\frac{1}{2}n^2$ for reactions between identical nuclei in a single gas, which is the number of possible reacting pairs per unit volume, and may be measured in cm^{-6}.

Thus r is proportional to the square of the density of the gas, for a mixture in given proportions.

If r is multiplied by the energy Q released in each reaction, the result is the total rate of release of energy per unit volume per second, an important parameter of any thermonuclear reaction.

Another important parameter is the probability per second that a given nucleus will undergo reaction. For a nucleus of type 1, this is

$$P = \frac{r}{n_1} = n_2\langle \sigma v \rangle, \qquad\qquad \textbf{10.9}$$

which is measured in reciprocal seconds. P is proportional to the number of nuclei of type 2 per unit volume, and hence to the density of the gas if the composition is constant.

The mean life of a nucleus of type 1 before it disappears by reaction is $\tau = P^{-1}$. The mean life at a given temperature and density is a convenient number for comparing one thermonuclear reaction with another. If n_2 is so much larger than n_1 that it is hardly changed by the reaction, the mean life of a single nucleus of type 1 is also the time taken for the number of such nuclei to fall by a factor e. They are in fact burnt up exponentially, the number remaining at time t being proportional to e^{-Pt}.

With reactions in a single gas, or between the components of a fifty–fifty mixture, the material is used up more slowly than exponentially, if the volume and temperature are kept constant.†

It should be noted that for reactions between identical nuclei in a single gas, P is given by

$$P = \frac{r}{n} = \tfrac{1}{2}n\langle\sigma v\rangle$$

instead of by equation 10.9.

10.3 Conditions for thermonuclear reaction

10.3.1 *Numerical calculations*

Calculations of the rates of individual reactions have been made by many authors, including those mentioned in section 10.2.2. A selection of results drawn from these sources, and from Gamow and Critchfield (1949), is shown in Figure 40, in terms of P, the probability per second of reaction of an individual nucleus, in gas mixtures containing one gramme-atom per cubic centimetre of each component. This is an arbitrary density, chosen for purposes of comparing reactions; it is higher than is likely to be practicable in the laboratory, but lower than is thought to hold in most stars.

It must be emphasized that in order to accommodate a lot of information the vertical scale of Figure 40 has been made extremely coarse, extending logarithmically over a total range of 10^{40}. For all practical purposes the top of the scale means 'unobservably fast', while the bottom means 'unobservably slow'.

In the following sections, the numbers shown in Figure 40 are used to support conclusions which could equally well have been drawn directly from equation 10.8.

10.3.2 *Temperature dependence*

Even on the logarithmic scale of Figure 40, all the reaction rates show steep temperature dependence between 10^6 and 10^8 K. Because the height of the potential barrier for given radius is proportional to $Z_1 Z_2$, the slope is smaller for reactions between the isotopes of hydrogen; but even here the mean life drops by a factor 10^{13} as we go from 10^6 K to 10^8 K. With the higher potential barrier of the reaction $^{14}N(p, \gamma)^{15}O$, the slope of the logarithmic plot is twice as great, giving a factor 10^{13} in reaction rate for a factor ten in temperature.

†As an exercise, show that for a d + d reaction at constant temperature and volume, starting with n_0 deuterons per cubic centimetre, the number remaining after time t would be

$$n = \frac{n_0}{P_0 t + 1},$$

where $P_0 = \tfrac{1}{2}n_0\langle\sigma v\rangle$.

158 Thermonuclear Reactions

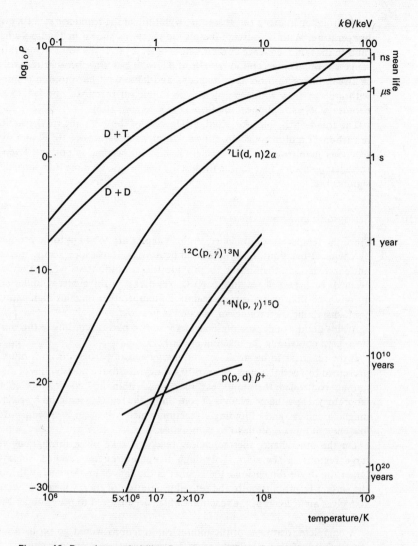

Figure 40 Reaction probability P and mean life τ, for density one gramme-atom per cubic centimetre for each reacting nucleus, i.e. one gramme-atom ^{7}Li + one gramme-atom d for ^{7}Li(d, n)2α or two gramme-atoms d for d+d

10.3.3   *Z-dependence*

It was pointed out in the preceding section that the product $Z_1 Z_2$ of the atomic numbers of the reacting nuclei fixes the height of the potential barrier and hence the slope of the plot of reaction rate against temperature. It is also

159   Conditions for thermonuclear reaction

a major factor in fixing the absolute magnitude of the reaction rate at a given temperature. With one exception, all the reactions shown in Figure 40 have rates in the same sequence as their values of $Z_1 Z_2$ over the whole temperature range shown. The exception is $^1\mathrm{H}(\mathrm{p}, \mathrm{d})\beta^+$, which is slow because it involves the weak interaction between nucleons and leptons; it has more in common with β-decay than with the general run of nuclear reactions, and is included because of its astrophysical importance (see section 10.5).

The trend of the plot for $^7\mathrm{Li}(\mathrm{d}, \mathrm{n})^4_2\mathrm{He}$, and of those for the (p, γ) reactions in carbon-12 and nitrogen-14, shows that at temperatures of 10^9 K and over, reactions involving the light nuclei ($Z = 3\text{--}10$) become important. Even at these temperatures, reactions in the heavier nuclei have no more than secondary significance.

10.3.4 *Self-sustaining thermonuclear reactions*

Since the temperatures discussed in this chapter are too high to be produced by chemical reactions, or by electrical heating, we must look to the thermonuclear reactions themselves to provide the conditions in which they can occur. If each pair of reacting particles releases a positive energy under conditions such that over a certain region the temperature remains high enough, a self-sustaining thermonuclear reaction is possible.

At this point, the discussion separates into two parts according to the size of the region containing the reacting material.

If the region is to be small, loss of energy and of particles from it must be prevented by special means. Section 10.4 is devoted to the possibilities of maintaining controlled thermonuclear reactions within limited volumes in the laboratory. Here the problem is of containing the reacting material by electromagnetic means, and of finding a reaction which will occur at a temperature low enough for containment to be possible.

On the other hand, thermonuclear reactions take place throughout very large regions in the stars. Very high temperatures are reached, since the larger the linear dimensions, the smaller is the rate of loss of energy through the surface in proportion to the rate of generation of energy inside it. Section 10.5 gives an account of the reactions which are believed to occur under these conditions.

A complete treatment of thermonuclear reactions would go on to discuss the synthesis of elements in the stars, by secondary reactions occurring in the extreme conditions which are maintained by the primary, self-sustaining reactions. The secondary processes can include reactions with negative Q, in which large nuclei are built up out of stabler medium-sized ones. Since that is a story far beyond the scope of this book, the reader is referred to the reviews of it by Burbridge *et al.* (1957) and Tayler (1966).

10.4 Controlled thermonuclear reactions

10.4.1 *Choice of reaction*

The choice of reaction for attempts at achieving a controlled thermonuclear reaction in the laboratory is indicated by Figure 40. At a temperature of 10^7K, the only serious candidates are the d + t reaction and the d + d reactions. The nearest competitor is ^{7}Li(d, n)$^4_2$He, which is 10^7 times slower.

At 10^7K and total density 2 gramme-atoms per cubic centimetre (i.e. $n = 2 \times 6 \cdot 023 \times 10^{23}$ cm^{-3}), P for the reaction ^{2}H(d, n)^{3}He is 53 s^{-1}, as shown on Figure 40.

A more realistic figure for the maximum density of plasma obtainable in the laboratory is $n = 10^{18}$ cm^{-3}, which would give $P = 4 \cdot 4 \times 10^{-5}$ s^{-1}, and a reaction rate $r = 4 \cdot 4 \times 10^{13}$ s^{-1} cm^{-3}.

With $Q = 3 \cdot 25$ MeV $= 5 \cdot 2 \times 10^{-13}$ joules,

this reaction rate gives a rate of generation of energy

$$rQ = 2 \cdot 3 \text{ W cm}^{-3}.$$

In fact we have also the reaction ^{2}H(d, p)^{3}H occurring at the same rate with $Q = 4 \cdot 03$ MeV and therefore giving a further release of energy at a rate of $2 \cdot 8$ W cm^{-3}. Thus the primary d + d reactions in deuterium at density $n = 10^{18}$ atoms cm^{-3} at a temperature of 10^7 K could be expected to release energy at a rate of $5 \cdot 1$ W cm^{-3}. However, the reaction products ^{3}H and ^{3}He would ultimately undergo further reactions yielding α-particles, with a total Q of $21 \cdot 6$ MeV per d + d reaction. The total rate of release of energy in deuterium at this density and temperature would therefore become $15 \cdot 2$ W cm^{-3}.

Corresponding figures for the d + t reaction ^{3}H(d, n)^{4}He at 10^7 K are:

$$Q = 17 \cdot 58 \text{ MeV} = 28 \cdot 1 \times 10^{-13} \text{ joules}.$$

For total density 2 gramme-atoms per cubic centimetre:

$$n_1 = n_2 = 6 \cdot 023 \times 10^{23} \text{ cm}^{-3},$$

$$P = 3 \cdot 33 \times 10^3 \text{ s}^{-1} \quad \text{(see Figure 40)}.$$

For total density 10^{18} atoms per cubic centimetre:

$$n_1 = n_2 = 5 \times 10^{17},$$

$$P = 2 \cdot 78 \times 10^{-3} \text{ s}^{-1},$$

$$r = 1 \cdot 39 \times 10^{15} \text{ s}^{-1} \text{ cm}^{-3},$$

$$rQ = 3900 \text{ W cm}^{-3}.$$

Thus if we could maintain a deuterium–tritium plasma at 10^7 K, it would generate power at a rate of $3 \cdot 9$ kW cm^{-3}, compared with 15 W cm^{-3} for the total yield from pure deuterium. Either of these can be imagined as the basis of a controlled source of thermonuclear power.

10.4.2 *Confinement of the plasma*

We come now to the problem of confining within a fixed volume a plasma at a temperature of several million degrees. Clearly a solid enclosure can do no more than act as an outer container with some other agency keeping the hot plasma away from the walls.

Nearly all the work done so far on plasmas suitable for thermonuclear reactions has used a magnetic field as the principal confining agent. The confining effect of a magnetic field on a plasma is a subject in itself, demanding specialized treatment especially for determining the nature of the instabilities which limit the usefulness of so many configurations (see, for example Linhart, 1960, and Thompson, 1964).

Here we shall simply point out that the effect of a steady magnetic field on a single moving charged particle is to rotate the velocity vector around the direction of the field, without changing the magnitude of the velocity. Thus a particle moving in the direction of the field is unaffected, but a particle with a component perpendicular to it follows a helical path, with constant component of velocity in the direction of the field if the latter is uniform. If the magnetic field is non-uniform, the path is part of a distorted helix and the particle is prevented from entering a region where the field exceeds a limit set by the incident velocity.

In a plasma, the net effect of the forces on the individual particles is to allow unimpeded current flow only along the lines of magnetic field $\mathbf{B}$, while any transverse current flow gives rise to lateral forces; these deflect particles in a direction perpendicular to $\mathbf{B}$ and to $\mathbf{j}$, until the sideways drift is halted by the pressure gradient resulting from local accumulation of particles. The total force on the particles in unit volume is made up of the electromagnetic force $\mathbf{j} \wedge \mathbf{B}$, and $(-\mathrm{grad}\ p)$ from the non-uniformity of pressure p. At equilibrium this total force is zero, whence

$$\mathrm{grad}\ p = \mathbf{j} \wedge \mathbf{B}.$$

This implies a concentration of the plasma in the region of lowest magnetic field as can be seen if we use Maxwell's equation for curl $\mathbf{B}$

$$\mathrm{curl}\ \mathbf{B} = 4\pi \mathbf{j}.$$

Eliminating $\mathbf{j}$, we get

$$4\pi\ \mathrm{grad}\ p = (\mathrm{curl}\ \mathbf{B}) \wedge \mathbf{B},$$

which is a vector in the direction of decreasing B.

Methods of using these principles may be classified according to whether they use (a) a field produced by currents in external conductors, (b) the field due to a current flowing through the plasma itself, or (c) a field generated by beams of other particles passing through or beside the plasma.

Among systems of type (a), two well-known configurations are the toroidal and the mirror schemes:

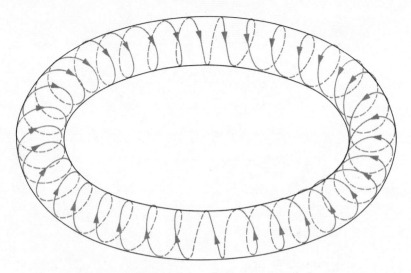

Figure 41 Toroidal magnetic field

The toroidal system, illustrated in Figure 41, appears at first sight to be ideal. But in fact the curvature of a simple toroidal field leads indirectly to a radial drift of the plasma towards the walls, and correcting devices are necessary. In one toroidal machine, the Stellarator, the toroid is bent to a figure eight, and each magnetic field line spirals round so that instead of making a single closed loop it generates a complete toroidal surface.

Another well-known toroidal machine, Zeta, comes partly into type (b) because the externally applied toroidal field is supplemented by the pinch effect, in which the magnetic field of a current in the plasma itself causes concentration of the ions into a region close to the axis of the toroid.

The mirror scheme, which belongs entirely to type (a), owes its name to the properties of a field of the shape shown in Figure 42(a). Particles going along the axis can leak through to the other side, but particles off the axis, or not moving parallel to it, are reflected backwards as they approach the region of high field. Two such mirrors, connected by a region of cylindrically symmetric field as shown in Figure 42(b), constitute a moderately efficient system of confinement, which is sometimes called a magnetic bottle. Various devices have been used for reducing the leakage of particles along the axis, which constitutes the neck of the bottle; for example, a radio-frequency field superposed on the main one can keep the bottle-necks moving, so that a trajectory which might have led to escape becomes one that leads to confinement before any particle has had time to use it.

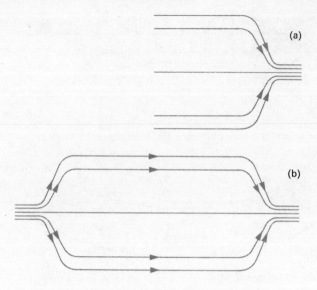

Figure 42 (a) Magnetic mirror. (b) Magnetic bottle

10.4.3 *Heating of the plasma*

To form a plasma, and raise it to the temperature needed for a thermonuclear reaction, initial heating is necessary. Though no scheme is pure, we may list the main heating processes as (a) acceleration of ions by the changing magnetic field during build-up of the confining field, followed by collisions in which directions and energies become random; (b) Joule heating by passage of current through the plasma; (c) acceleration of ions in a parallel beam which can be injected into an incomplete magnetic bottle and then trapped by switching on the rest of the confining field. After the bottle is closed, the directions and energies are quickly made random by collisions with the confining field or with other ions.

10.4.4 *Prospects*

In spite of massive research effort, distributed throughout the world, no controlled thermonuclear reaction has yet been achieved. However, great progress has been made in understanding the various possible schemes for confining and heating suitable plasmas.

In 1958 neutrons were observed from the toroidal machine Zeta at Harwell, operating with deuterium at about 10^6 K. Measurement of the spectra of neutrons emitted in different directions showed that they were being produced predominantly in collisions of deuterons moving parallel to the axis of the toroid and not in randomly directed collisions. The reactions were therefore declared to be nuclear reactions between artificially accelerated deuterons, not genuine thermonuclear reactions.

At the Culham laboratory of the UK Atomic Energy Authority, magnetic mirror and bottle systems have received thorough investigation in experiments with Thetatron and Phoenix devices; the difficulty of eliminating losses from these open-ended systems has caused interest to be concentrated on toroidal systems.

In the USA and the USSR so many research programmes are proceeding that the only reliable way of summarizing the world situation is to say that a summary would be out of date before it was written, let alone read. It is reasonable to expect that major progress towards the controlled release of thermonuclear energy will be reported in the not-very-distant future; but which techniques and which laboratories will figure in these reports of success, or indeed in the subsequent development of practical power sources, is still a matter of pure guesswork – unprofitable guesswork since success will in fact depend upon the whole body of understanding built up by all the techniques and all the laboratories.

10.5 Thermonuclear reactions in stars

10.5.1 *Review*

As has been mentioned already, a star provides ideal conditions for the occurrence of self-sustaining thermonuclear reactions. Because loss of plasma from the surface is inhibited by the gravitational field, through which there is no leakage, and the volumes are so large, the ranges of time, temperature and density are virtually unlimited in comparison with those that might be achieved on earth.

The formation of a star appears to start with a localized fluctuation in the density of the interstellar hydrogen gas; any local increase of density tends to grow as gravitational attraction brings in more hydrogen. As the mass increases, the pressure and density at the centre increase and the gravitational energy released by the contraction raises the gas to a high temperature. When the centre becomes hot enough and dense enough, thermonuclear reactions start to release more energy and convert hydrogen into helium by the processes described in section 10.5.2.

At a later stage, when the temperature is higher and there is more helium present, there is a possibility of formation of carbon nuclei via the short-lived beryllium-8 nucleus; thereafter, indirect formation of more helium may occur through the carbon–nitrogen cycle (see section 10.5.3), which is thought to predominate in stars hotter than the sun.

10.5.2 *The proton–proton chain*

The formation of alpha particles from free protons involves several steps, of which the first limits the over-all rate of the process. This is the reaction

$$^1_1H + ^1_1H \longrightarrow ^2_1H + \beta^+ + \nu \quad (Q = 0{\cdot}42 \text{ MeV}),$$

for which at 10^7 K and density 2 g cm^{-3} the mean life of a proton is 10^{14} years (see Figure 40).

The deuterons formed by this process have a mean life of only a few hours for formation of helium-3 by the reaction

$$^1_1H + ^2_1H \longrightarrow ^3_2He + \gamma \quad (Q = 5{\cdot}49 \text{ MeV}).$$

The final process in the chain is the formation of an alpha particle by the reaction

$$^3_2He + ^3_2He \longrightarrow ^4_2He + 2^1_1H \quad (Q = 12{\cdot}8 \text{ MeV}).$$

The total energy release in this process, with the two $p + p$ reactions and two $p + d$ reactions that must precede it, is $26{\cdot}7$ MeV; the net effect is formation of one alpha particle from four free protons.

10.5.3   *The carbon–nitrogen cycle*

As soon as there is a reasonable concentration of helium present, beryllium-8 nuclei can be formed either by collision of two alpha particles:

$$^4_2He + ^4_2He \longrightarrow ^8_4Be + \gamma,$$

or by the sequence of reactions,

$$^4_2He + ^3_2He \longrightarrow ^7_4Be + \gamma,$$

$$^7_4Be + ^1_1H \longrightarrow ^8_5B + \gamma,$$

$$^8_5B \longrightarrow ^8_4Be + \beta^+ + \nu.$$

Beryllium-8, however, has a mean life of only 10^{-15} s for break-up into two alpha particles, and the sequence will stop at this point unless the density of helium is enough for this to have a reasonable chance of being forestalled by a collision giving

$$^8_4Be + ^4_2He \longrightarrow ^{12}_6C + \gamma.$$

Once this link has been established, and a significant number of $^{12}_6C$ nuclei formed, further liberation of energy may occur by the following sequence of reactions:

$$^{12}_6C + ^1_1H \longrightarrow ^{13}_7N + \gamma,$$

$$^{13}_7N \longrightarrow ^{13}_6C + \beta^+ + \nu,$$

$$^{13}_6C + ^1_1H \longrightarrow ^{14}_7N + \gamma,$$

$$^{14}_7N + ^1_1H \longrightarrow ^{15}_8O + \gamma,$$

$$^{15}_8O \longrightarrow ^{15}_7N + \beta^+ + \nu,$$

$$^{15}_7N + ^1_1H \longrightarrow ^{12}_6C + ^4_2He.$$

This sequence is called the carbon–nitrogen cycle, and its net effect is the fusion of four free protons into an alpha particle, the carbon being regenerated after conversion to nitrogen. This indirect method of forming helium is important at high temperatures, because its rate is limited by that of the reaction $^{14}N(p, \gamma)^{15}O$, which although slower than $^{1}H(p, \beta^{+})^{2}H$ at 5×10^{6} K is much faster at 10^{8} K (see Figure 40).

Further reading

Review articles
R. Post, *Reviews of Modern Physics*, vol. 28, 1956, pp. 338–62.
R. Post, *Annual Review of Nuclear Science*, vol. 9, 1957, pp. 367–436.

Collections of papers on special topics, including Stellarator and Zeta programmes.
Longmore *et al.* (eds.), *Progress in Nuclear Energy*, series 11, *Plasma Physics and Thermonuclear Research*, vol. 1, 1959, Pergamon.

Chapter 11
Nuclear Forces

11.1   The nature of nuclear forces

11.1.1   *The problems*

'Nuclear force', in the present context, means the strong interaction that binds neutrons and protons together to form nuclei. We are not going to consider the weak and electromagnetic interactions which are responsible for β-decay and γ-decay of one nuclear state to another. Nor are we going to probe inside the neutron and the proton themselves; such probing comes under the heading of subnuclear physics, or elementary particle physics (see Hughes, 1971).

When we set about describing a nucleus in terms of the interactions between the neutrons and protons of which it is made up, we find ourselves facing many new problems, including the following three:

(a) How are the proton–proton, proton–neutron and neutron–neutron forces related to each other? Is the collective name nucleon (for neutron or proton) anything more than a verbal convenience?

(b) Is the force between two nucleons alone the same as the force between two nucleons in a nucleus?

(c) How must our quantum-mechanical methods be improved to handle a situation in which a particle is moving, not in a fixed central field, but in the collective field of a not very large number of other nucleons?

Problem (a) is discussed in section 11.2, where evidence is presented for the charge independence of nuclear forces, leading to the idea of treating the neutron and the proton as two charge-states of a single particle, the nucleon.

To answer problem (b) is part of the whole task of describing the nucleon–nucleon force. One meets it by saying that the force between two nucleons is determined by the states they occupy, and is dependent on other nucleons only in so far as these limit the states which the two can occupy.

However, different aspects of the nucleon–nucleon force are important in the two situations. In the nucleus one can describe the state of an individual nucleon moving in the collective field of the others; including spin–orbit interaction, and, symmetrizing the wave functions to allow for the indistinguishability of identical nucleons, one reaches the shell model of the nucleus (see Reid, 1972). In the interaction of two nucleons alone, leading to scattering or to the formation of a deuteron, the details of the nucleon–nucleon force show up better, and can be related more directly to measurable quantities. The neutron–proton force is discussed in detail in sections 11.3 and 11.4, with further discussion of its spin dependence in Chapter 12.

Problem (c) is answered by hard work and electronics. The adaptation of simple quantum mechanics to the more complicated need is largely a matter of laborious computation. But many shell-model calculations which would formerly have had to be left undone, or covered by approximation, can now be carried out satisfactorily with the aid of electronic computers.

11.1.2 *The facts to be explained*

Among all the properties of individual types of nucleus, two generalizations stand out as requiring explanation. These are:

(a) The radii of nuclei with different values of the mass number A are roughly proportional to $A^{\frac{1}{3}}$. This means that the density of nuclear matter does not vary much with the size of the nucleus, and that nucleons are packed equally tightly in nuclei of all sizes. Whatever the nature of nuclear forces, it must explain the tendency of nucleons to pack with their centres about $1 \cdot 3 \times 10^{-13}$ cm apart.

(b) The binding energy per nucleon is between 7 and 9 MeV for all nuclei from carbon to uranium (see Figure 34, p. 135). This is strong evidence for the saturation of nuclear forces, by which we mean that each nucleon interacts only with a small number of its nearest neighbours. If it interacted with all the other nucleons in the nucleus, the total binding energy would be proportional to the total number of pairs of nucleons, which is $\frac{1}{2}A(A-1)$, and the binding energy per nucleon would be proportional to $A-1$.

11.1.3 *The nucleon–nucleon force*

Both the uniform density of nuclear matter and the saturation of nuclear forces suggest strongly that the force between two nucleons, instead of falling off with distance according to a simple power law, has a more complicated radial dependence involving a characteristic distance which may be called the range of the force. In fact it must be a short-range force.

If the force is independent of the spin orientation of the nucleons, and is directed along their line of centres, it is called a central force and may be described as the gradient of a potential. The plot of potential as a function of radius r is a convenient way of summarizing the important features of the force. For an attractive short-range force, this plot goes to negative values at small

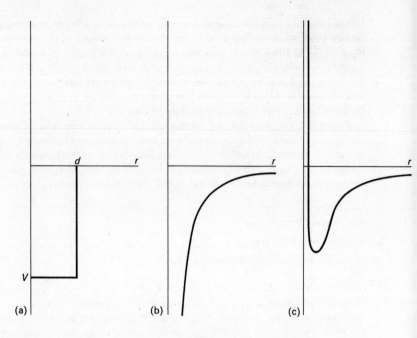

Figure 43 Potential wells: (a) square, (b) Yukawa, (c) repulsive core

values of r, and its appearance gives rise to the term 'potential well'. Three potential wells are illustrated in Figure 43; the simplest for many purposes is the square well (Figure 43a), in which the nucleons are bound together with energy V if their centres are closer than a distance d, but completely free if they are farther apart. Of course, this shape of well can only be used as an approximation, since it gives infinite force between the nucleons at $r = d$, and zero force otherwise. No mechanism is likely to come very close to providing such a force.

An attractive force of limiting range is provided by the Yukawa well (Figure 43b), in which the potential is given by

$$\phi = \frac{ge^{-kr}}{r}. \qquad \textbf{11.1}$$

The force, which is everywhere attractive, falls off rather faster than an exponential, and k^{-1} may be taken as the range.

Without further assumption, neither the square well nor the Yukawa well explains the constant density of nuclear matter. Either of them, in the simple form described above, would suggest that a nucleus was stablest when collapsed to a very small radius. However, it may be shown (Blatt and Weisskopf, 1952, pp. 128–33) that constant density of nuclear matter, and saturation of nuclear forces, are both obtained if the short-range attractive force between pairs of nucleons with symmetric wave functions is balanced by an equally strong repulsive force between pairs of nucleons in antisymmetric states. This requires the force to be dependent on the particular state occupied by the nucleons (see section 11.1.4 on exchange forces).

An alternative and more direct way of getting constant density is to equip each nucleon with a repulsive core. This gives it a potential of the shape shown in Figure 43(c), a well with a tall thin tower rising from its bottom. While such a well may look improbable, it is a simple empirical device using a central potential to prevent the collapse of the nucleus that would occur under a purely attractive nuclear potential. There is some evidence, from high-energy nucleon–nucleon scattering, for a type of core structure, but the question is still open, and study of nuclear forces continues without any certainty about what happens to the nucleon–nucleon force at very short range.

11.1.4 *Exchange forces*

We have discussed the radial dependence of the nucleon–nucleon force, from the point of view of the facts which have to be explained. Now it is time to think about the mechanism of the force itself.

In the very early days of nuclear theory, Heisenberg (1932) turned to exchange forces for the source of a short-range nucleon–nucleon force, of sign depending on the state occupied by the two nucleons. The crudest picture of an exchange force involves the passing of an intermediate particle from one to the other of the interacting nucleons. This idea develops, as we learn that the exchanged particle can be virtual instead of real (i.e. it can have momentum and energy incompatible with its appearance as a free particle). Then we find that the mere exchange of a set of quantum numbers is enough. In the formal description, we specify an exchange force by an operator which acts on the wave function of the system to produce the wave function which it would have after the exchange process.

For example, in Heisenberg exchange, we imagine the force as resulting from the exchange of identities of two indistinguishable particles; that is, particle A is replaced by particle B, and vice versa. As an alternative description, we can say that the two particles exchange spin coordinates and space coordinates. Then if $\psi(r_1, s_1; r_2, s_2)$ means the wave function of the system in which nucleon A has position r_1 and spin s_1, while nucleon B has position r_2 and spin s_2, the Heisenberg operator P_H is defined by

$$P_H \psi(r_1, s_1; r_2, s_2) = \psi(r_2, s_2; r_1, s_1).$$

It is now considered that the main components of the nucleon–nucleon force come from Heisenberg exchange and from Majorana exchange. In the latter, positions are exchanged, but not spins. The Majorana exchange operator is therefore defined by

$$P_m \psi(r_1, s_1; r_2 \, s_2) = \psi(r_2, s_1; r_1, s_2).$$

A third type of exchange force is defined by the Bartlett exchange operator which interchanges spins but not positions, as follows:

$$P_B \psi(r_1, s_1; r_2, s_2) = \psi(r_1, s_2; r_2, s_1).$$

The list of possible short-range forces is now concluded with the Wigner force, which is an ordinary or non-exchange type of short-range force.

11.1.5 *Meson theory*

The Yukawa potential is important because simple arguments show that a short-range force described by a potential of this shape should have associated with it particles of rest mass

$$m = \frac{k\hbar}{c}, \qquad \qquad \textbf{11.2}$$

where k is the constant in equation **11.1**. In fact the rest mass is such that the range of the force is

$$k^{-1} = \frac{\hbar}{mc},$$

which is the Compton wavelength of the particles.

These particles, which may be considered as the quanta of the nuclear force, were suggested by Yukawa (1935), observed by Perkins (1947) and by Occhialini and Powell (1947). They are now known as pions, or pi mesons, having been distinguished from the weakly interacting mu mesons by Lattes, Occhialini and Powell (1947).

The elementary arguments of meson theory, leading to equation **11.2**, are summarized in many books, including Hughes (1971). They are therefore not repeated here.

Free pions exist in three charge-states; those with positive or negative charge have mean life 2×10^{-8} s, but the neutral pions have a much shorter life of about 10^{-16} s.

The exchange forces between a neutron and a proton may be thought of as involving the exchange of a virtual charged pion, while neutron–neutron or proton–proton forces involve exchange of neutral virtual pions.

11.2 Charge independence

11.2.1 *Charge symmetry*

The question to be considered now is problem (a) of section 11.1: how is the nuclear force between a neutron and a neutron related to that between a neutron and a proton, and to that between a proton and a proton? We limit the question to the nuclear force to eliminate the obvious difference that there is a Coulomb repulsion between two protons.

A start may be made by studying the binding energies of pairs of mirror nuclei with odd A, for example ^7_3Li and ^7_4Be, $^{11}_5\text{B}$ and $^{11}_6\text{C}$, or $^{13}_6\text{C}$ and $^{13}_7\text{N}$. Mirror nuclei are pairs of nuclei related by the interchange of protons with neutrons. In other words, they contain equal numbers of protons and neutrons, plus one (or more) extra which is either a neutron or a proton. For example $^{11}_5\text{B}$ and $^{11}_6\text{C}$ consist of five neutrons and five protons, plus one extra neutron for $^{11}_5\text{B}$ or one extra proton for $^{11}_6\text{C}$. $^{11}_6\text{C}$ will have extra Coulomb energy which may be calculated from the work done in bringing an extra proton up to a nucleus already containing five protons; the result depends somewhat on the assumptions made about the distribution of charge in the nucleus.

If Z protons, each of charge e, are distributed over the volume of a spherical nucleus of radius R, their Coulomb energy is

$$\frac{3}{5}\frac{Z(Z-1)e^2}{R}. \qquad\qquad \textbf{11.3}$$

The difference between the Coulomb energies of two mirror nuclei, assumed to have the same radius, one containing Z protons and $N(=Z-1)$ neutrons, the other $Z-1$ protons and $N+1(=Z)$ neutrons, would therefore be

$$\frac{6}{5}\frac{(Z-1)e^2}{R}. \qquad\qquad \textbf{11.4}$$

To get the total expected difference between the masses of the two mirror nuclei, we must include the difference $m_\text{p}-m_\text{n}(=-0\cdot78\text{ MeV})$ of the masses which the extra proton and neutron would have in the free state. (Remember that all these masses refer to neutral atoms.)

Any difference between the quantity

$$\frac{6}{5}\frac{(Z-1)e^2}{R}-0\cdot78\text{ MeV} \qquad\qquad \textbf{11.5}$$

and the observed mass difference should then represent a difference in the energies of interaction of the extra neutron and proton with the rest of the nucleus.†

---

†The expressions **11.3-5** assume that each proton exists in the nucleus as a sphere of the same radius as a free proton. The Coulomb self-energy of each proton is ignored. But if the charge were distributed continuously, instead of in units of e on proton-sized spheres, the total Coulomb energy would be

$$\frac{3}{5}\frac{Z^2e^2}{R},$$

and the difference in the Coulomb energies of two mirror nuclei would be

$$\frac{6}{5}\frac{(Z-\frac{1}{2})e^2}{R}.$$

Then one would have to allow for the part of m_p which represents the Coulomb self-energy of the free proton. The net result would differ from expression **11.9** by the Coulomb energy released when a charge e, formerly distributed over a proton, redistributes itself over a larger nucleus. This is

Table 12 Comparison of Masses of Mirror Nuclei, Illustrating Charge Symmetry

Observed mass/a.m.u.	7·016 929 ($^7_4$Be)	11·011 433 ($^{11}_6$C)	13·005 739 ($^{13}_7$N)
	7·016 005 ($^7_3$Li)	11·009 305 ($^{11}_5$B)	13·003 354 ($^{13}_6$C)
Observed mass /a.m.u.	0·000 924	0·002 128	0·002 385
difference /MeV	0·86	1·98	2·22
Difference in Coulomb energy /MeV −0·78	1·02	1·80	2·15
Discrepancy MeV	+0·16	−0·18	−0·07
Total interaction energy of extra neutron/MeV	7·25	11·45	4·95

Numerical comparisons, for the three pairs of mirror nuclei mentioned above, are made in Table 12. The discrepancies obtained are small enough to show that the observed mass differences agree with those calculated from Coulomb energies, within the expected accuracy of the latter. While this does not prove that the interaction energy of the extra neutron with the rest of the nucleus is equal to that of the extra proton with the same 'rest of nucleus', it does make clear that any genuine discrepancy is much smaller than the inter-action energy itself. The latter quantity is listed in the last line of the table, for the extra neutron. For example, the interaction energy of n + $^6_3$Li to make $^7_3$Li is 7·25 MeV, while the non-Coulomb interaction energy of p + $^6_3$Li to make $^7_4$Be is calculated as 7·41 MeV.

The facts listed in Table 12 do not stand alone. They form a small part of a great volume of information about binding energies of extra nucleons, and about excited states. For example, $^{11}_5$B and $^{11}_6$C are similar not only in their binding energies but also in their excited states. In the summary by Ajzenberg-Selove and Lauritsen (1959), the energies of the first three excited states of $^{11}_5$B are

$$\frac{3}{5}(Z-1)e^2\left(\frac{1}{R_{\text{proton}}} - \frac{1}{R_{\text{nucleus}}}\right),$$

which amounts to 0·12 MeV for $A \simeq 11$.

This calculation may be taken as indicating that uncertainties in the details of the charge distri-bution lead to uncertainties of order 0·1 MeV in the Coulomb energy differences. The choice of the radius leads to a more important uncertainty ($\pm$0·2 MeV) in the results shown in Table 12.

given as 2·13, 4·46 and 5·03 MeV, while those of $^{11}_{6}C$ are at 1·99, 4·26 and 4·77 MeV. This close correspondence suggests that the nucleon–nucleon interactions in the two nuclei are very similar, and in fact the excited states give closer agreement than do the ground states in a comparison of the type shown in Table 12.

A moment's thought is needed to show what can, and what cannot, be deduced from the above facts. The total interaction energy of a neutron with $Z-1$ protons and $N(= Z-1)$ neutrons, i.e. of $(Z-1)$ n+p pairs and $(Z-1)$ n+n pairs, appears to be the same as that of a proton with $Z-1$ protons and $Z-1$ neutrons, that is, of $(Z-1)$ p+p pairs and $(Z-1)$ n+p pairs. Both of these contain the same number of n+p pairs, so we have learned nothing about the n+p interaction. The conclusion towards which our information points is that the neutron–neutron interaction is the same as the non-Coulomb part of the proton–proton interaction.

This equivalence goes by the name of *charge symmetry*. It implies that the nuclear forces are unaffected if all the neutrons are replaced by protons, and all the protons by neutrons. It is a part of the wider concept of charge independence which, as will be seen in the next section, brings in n+p forces also.

11.2.2 *Charge independence in light nuclei*

The idea of charge independence goes beyond the equivalence of n+n and p+p interactions, suggesting that the n+p interaction may be the same also, for nucleons in corresponding states.

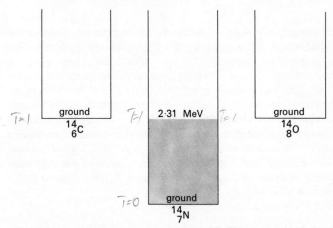

Figure 44 Energy levels of carbon-14, nitrogen-14 and oxygen-14, corrected for Coulomb effects

Mirror nuclei with odd A give no evidence on this point, so we move on to a less well-known pair of mirror nuclei with even A and two extra nucleons, namely $^{14}_{6}C$ with a core of $^{12}_{6}C$ and two extra neutrons, and $^{14}_{8}O$ with the same core and two extra protons. Here again, correction for the extra Coulomb energy of $^{14}_{8}O$, and for the difference $2(m_p - m_n)$ gives evidence for charge symmetry. But $^{14}_{7}N$, with the same core and a neutron–proton pair, has a lower mass even after the Coulomb correction (see Figure 44). This means that $^{14}_{7}N$ in its ground state has stronger nuclear binding forces than $^{14}_{6}C$ or $^{14}_{8}O$. We therefore explore the idea that the neutron–proton pair in $^{14}_{7}N$ may be in a state which is basically different from that of the two protons in $^{14}_{8}O$, or that of the two neutrons in $^{14}_{6}C$.

Before moving on, however, we note that the discrepancy is about 2·3 MeV, and that $^{14}_{7}N$ has an excited state at 2·31 MeV. This state is formed, along with the ground state and many other excited states, in inelastic scattering of protons, and in the reaction $^{13}_{6}C(d, n)^{14}_{7}N$, but it is peculiar in not being formed in the reaction $^{16}_{8}O(d, \alpha)^{14}_{7}N$, except weakly at high deuteron energies (see Noble, 1969, and also Freemantle, et al. 1953).

These facts are explained by saying that in the ground-state of $^{14}_{7}N$, the extra neutron and proton are in a state not accessible to the two neutrons of $^{14}_{6}C$ or to the two protons of $^{14}_{8}O$. But in the 2·31 MeV excited state of $^{14}_{7}N$, the extra neutron and proton are in the state corresponding to that of the two neutrons in $^{14}_{6}C$, and of the two protons in $^{14}_{8}O$. It is thus the central member of a triplet of states, in which the nucleon–nucleon force is the same whether it is $n + n$, $n + p$ or $p + p$. The energies are plotted in Figure 44.

11.2.3 Isospin

In 1932 Heisenberg put forward a scheme for treating the neutron and the proton as nucleons with different values of a charge quantum number. This scheme was extended by Wigner (1937), who named the charge quantum number isotopic spin and showed how it could be developed along lines closely similar to those used for ordinary spins. The importance of the concept increased with its application to light nuclei (see, for example, Inglis, 1953), and the name was changed to isobaric spin, to rectify the anomaly that it related isobars, not isotopes. Indeed it has been suggested that isotopic spin was so named because it was neither isotopic nor a spin! Now, with a great increase of its use in the description of fundamental particles, the name has been contracted to isospin or i-spin. We use for it the symbol T; some authors call it I, if confusion with other uses for this letter is unlikely.

The sequence of ideas is as follows:

(a) If the proton and the neutron are equivalent in their nuclear interactions, we should consider them as two states of a single particle, the nucleon, the two states being distinguished by two alternative values of a charge quantum number.

(b) Another well-known quantum number which has two permitted alternative values is the z-component of angular momentum of a particle of spin $\frac{1}{2}$. Let us give the nucleon an isospin T of $\frac{1}{2}$, which can have a z-component T_3 of $\frac{1}{2}$ or $-\frac{1}{2}$. When $T_3 = \frac{1}{2}$, the nucleon is a proton; when $T_3 = -\frac{1}{2}$, it is a neutron. The vector T and its z-component T_3 are analogous to an ordinary spin and its z-component, and it is this analogy that gives rise to the word 'spin' in all the names given to T. But T and T_3 exist in a hypothetical isospin space which has no connexion with real space, and none with angular momentum. The external agencies for altering and observing the value of T_3 have to do with electric charge.

(c) Systems of several particles will have total isospins which are related to those of the individual particles by the rules which hold for vector addition of ordinary quantum-mechanical angular momenta. For example, two nucleons each with $T = \frac{1}{2}$ can have a total T of 0 or 1; if $T = 0$, T_3 can only be 0, and the nucleons must be a neutron and a proton; but if $T = 1$, T_3 can be $+1$ (representing two protons), or 0 (representing a neutron and a proton), or -1 (representing two neutrons).

(d) If nuclear forces are charge independent, they will be the same in systems of given T but different T_3. The main properties of a system will be defined by its value of T, while T_3 merely serves to distinguish different charge-states of the system.

(e) Conservation of T_3 follows from conservation of charge and of number of nucleons, in processes which do not involve strange particles (see Hughes, 1971).

(f) Charge independence of nuclear forces means invariance of nuclear forces forces under changes from one charge-state to another. Such changes may be described as rotations in isospin space. Invariance of energy under rotations in ordinary space leads to conservation of angular momentum. Correspondingly, invariance of nuclear forces under rotations in isospin space leads to conservation of the isospin T.

(g) At the beginning of this chapter we pointed out that the nuclear forces to be discussed would be the strong nuclear interactions, not the weak and electromagnetic interactions. It is only the strong interaction that is charge independent. Consequently we expect the isospin T to be conserved in processes which go by the strong interaction; this means ordinary nuclear reactions which take place in characteristic times of order 10^{-22} s. Isospin is not expected to be conserved in beta decays, which go by the weak interaction, nor in any processes involving electromagnetic radiation.

Now let us apply these ideas to the three nuclei $^{14}_{6}C$, $^{14}_{7}N$ and $^{14}_{8}O$ mentioned in the preceding section. As a start, we assume that the common core, $^{12}_{6}C$, has $T = 0$ and $T_3 = 0$. Then the two extra neutrons in $^{14}_{6}C$ will have total $T_3 = -1$, and the two extra protons in $^{14}_{8}O$ will have total $T_3 = +1$. Consequently they must be in a state with $T = 1$. An isospin $T = 1$ can have three possible values of its z-component T_3, namely $1, 0$ and -1. The level at 2·31 MeV in $^{14}_7$N, with nuclear binding energy equivalent to that of the ground states of $^{14}_6$C and $^{14}_8$O, may therefore have $T = 1$ and be the centre member of the isospin triplet of which the other members are $^{14}_6$C and $^{14}_8$O. The ground state of ^{14}N is then an isospin singlet, with $T = 0$, the two extra nucleons being in a state which cannot contain two nucleons of the same charge. It appears that this state is more tightly bound than the triplet with $T = 1$.

The assignment of $T = 1$ for the excited state of $^{14}_7$N at 2·31 MeV is borne out by the fact that this state is not formed readily in the reaction $^{16}_8$O(d, α)$^{14}_7$N. The nuclei $^{16}_8$O, d and α all have $T = 0$, and conservation of T in this reaction limits the $^{14}_7$N to states with $T = 0$. On the other hand, in the reaction $^{13}_6$C(d, n)$^{14}_7$N, and in elastic scattering of protons by $^{14}_7$N, vector addition of isospins shows that $^{14}_7$N can be formed in states with either $T = 1$ or $T = 0$.

11.2.4 *The generalized Pauli principle*

The Pauli exclusion principle is usually quoted as saying that two identical fermions (i.e. particles with half-integral spin, which obey Fermi–Dirac statistics) cannot occupy the same state. However, when we have two identical particles with spin one half, in states with values $+\frac{1}{2}$ and $-\frac{1}{2}$ for the individual values of the z-component of angular momentum (m), we cannot say which particle has which value of m, and it is better to speak of the state of the whole system, with two particles jointly occupying the two substates. The Pauli principle is then equivalent to the following rule:

Pairs of identical fermions can exist only in states whose total wave functions are antisymmetric (i.e. change sign) under interchange of the two particles.

The total wave function is the product of two parts which we may write as $\phi(r_1, r_2)$ dependent on the space coordinates of the two particles, and $\chi(S; m_1, m_2)$, dependent on the z-components of their spins, and on the total spin S. The quantity $m_1 + m_2$ is the z-component of S.

When we interchange the two particles, ϕ will change sign if the particles are in a state of motion with odd orbital angular momentum l; but if the particles have no relative motion, or have motion with an even value of l, ϕ is unaffected by the interchange.

On the other hand, χ cannot be affected by interchanging the particles if they have the same value of m. If $m_1 \neq m_2$, χ may change sign or not. In fact it changes sign for states with total spin $S = 0$, and remains unaffected for states with $S = 1$, for any of the three possible values $(1, 0, -1)$ of $m_1 + m_2$, which is the z-component of S.

The total wave function

$$\psi = \phi\chi$$

is antisymmetric if ϕ or χ, but not both, change sign on interchanging of the particles.

Hence the new formulation of the Pauli principle includes the old, by stating that a pair of identical fermions with zero orbital angular momentum of relative motion can exist only in states with zero z-component of total spin. It goes farther, saying that the total spin itself must be zero, that is, the state $S = 1$, $m_1 + m_2 = 0$ is forbidden. It also tells us that in states with non-zero orbital angular momentum, S must be 1 if l is odd, or 0 if l is even. Everything is summed up in the condition

$l + S =$ even.

Now we may consider the generalization of the Pauli principle to cover nucleons which may be in different charge states. A neutron and a proton may not be identical in the sense used above, but in the charge-independent picture they may be considered as identical nucleons with different charge quantum numbers, that is, identical particles in different states. So we build up the total wave function of a pair of nucleons out of three parts, $\phi(r_1, r_2)$ and $\chi(S; m_1, m_2)$ as above, and a third part $\tau(T; T_{31}, T_{32})$ which depends on the total isospin T and on the z-components of isospin for the two nucleons (T_{31} and T_{32}). This isospin wave function is closely analogous to the spin wave function χ: it changes sign on interchange of the particles if $T = 0$, but not if $T = 1$.

The total wave function, which now includes the isospin part τ and may be written

$\psi = \phi\chi\tau$

is therefore still antisymmetric for the permitted states of a pair of protons or a pair of neutrons. So we are including the ordinary Pauli principle if we propose, as generalized Pauli principle, the rule that:

The total wave function, including space, spin and isospin parts, must be antisymmetric for a pair of nucleons.

This gives us restrictions on the possible states of a neutron–proton pair. For example, in a state with $T = 1$, their spin wave function χ must be anti-symmetric, with $S = 0$, provided that the energy is low enough to limit them to $l = 0$. Thus the 2·31 MeV excited state of $^{14}_{7}$N should have ordinary spin zero, along with the ground states of $^{14}_{6}$C and $^{14}_{8}$O. But a neutron–proton pair in a state with $T = 0$, with isospin wave function τ antisymmetric, should have spin wave function χ symmetric and hence total spin $S = 1$. This predicts that the ground state of $^{14}_{7}$N should have spin S equal to 1, a fact which is confirmed by spectroscopic observations (Townes *et al.*, 1947).

In terms of the orbital angular momentum l, the total spin S, and the total isospin T, the permitted states for a pair of nucleons are therefore given by

$l + S + T =$ odd.

11.3 The ground state of the deuteron

As a first exercise in fitting numerical values to the parameters of the nuclear force we consider the interaction between one neutron and one proton in the bound state which constitutes the deuteron.

11.3.1 *The data*

So far as is now known, the deuteron has no excited states, so the term 'ground state' in the title of this section is an indicator of fact rather than of selection.

The basic experimental data about the deuteron are that it has binding energy

$\varepsilon = 2 \cdot 225$ MeV,

nuclear spin

$I = 1$

and magnetic moment 0·857 nuclear magnetons.

The magnetic moment is practically equal to the sum of the separate magnetic moments of the proton and the neutron (see section 12.1), and with the value of I serves to tell us that the proton and neutron are predominantly in a $^3S_1$ state with their spins parallel, when they are bound together in a deuteron. Thus the deductions which we shall make from the value of the binding energy refer to the neutron–proton force in the triplet state with total spin one and zero orbital angular momentum.

11.3.2 *Calculation*

If the interaction of the neutron and proton in this state is described by a spherically symmetric potential $V(r)$, we may describe their relative motion by a spherically symmetric wave function $\psi(r)$, or by a modified wave function

$u(r) = r\,\psi(r).$

The function $u(r)$ satisfies an equation resembling the one-dimensional Schrödinger equation

$$\frac{d^2u}{dr^2} + \frac{2m}{\hbar^2}(E - V)u = 0, \qquad\qquad \textbf{11.6}$$

where E is the total energy, on a scale which makes $E - V$ the kinetic energy, and m is the reduced mass, approximately equal to half the nucleon mass (see equation **3.18**). The quantity r is the distance between the neutron and the proton, which is more a diameter than a radius.

The method of solving this equation depends upon the form assumed for the potential $V(r)$. Blatt and Weisskopf (1952, pp. 49–56) give details of how to compare results for various shapes of potential well (square, exponential, Gaussian and Yukawa), while Segré (1964, Appendix E) presents the 'effective-range' method of summing up neutron–proton scattering results in a form independent of well shape. However, as the simplest way of demonstrating the principles behind what is going on, we shall present only the calculation for a square well of radius d and depth V_t (see Figure 45).

With this square well, we must use the following values of V:

inside the well, where $r < d$, $V = -V_t$,

outside the well, where $r > d$, $V = 0$.

The total energy E is $-\varepsilon$ everywhere. This makes the kinetic energy inside the well

$$E - V = V_t - \varepsilon, \quad \text{which is positive.}$$

According to classical mechanics, values of $r > d$ would be impossible, for the kinetic energy outside the well would be

$$E - V = -\varepsilon.$$

Quantum-mechanically, however, some penetration into this region can occur. Just how much is found by solving equation **11.6** with $E - V = -\varepsilon$ and fitting the solution to that which holds inside the well, as follows.

Inside the well, the solution for u is oscillatory, of form

$$u(r) = B \sin Kr,$$

where $K^2 = \dfrac{2m}{\hbar^2}(V_t - \varepsilon).$ **11.7**

But outside, there is an exponential decrease, given by

$$u(r) = A e^{-\alpha r},$$

where $\alpha^2 = \dfrac{2m}{\hbar^2}\varepsilon.$ **11.8**

The continuity of u across the boundary at $r = d$ requires

$$A e^{-\alpha d} = B \sin Kd$$ **11.9**

and the continuity of du/dr requires

$$-\alpha A e^{-\alpha d} = KB \cos Kd.$$ **11.10**

Dividing equation **11.10** by equation **11.9** gives

$$K \cot Kd = -\alpha.$$ **11.11**

Since K and α are given by equations **11.7** and **11.8** in terms of the well depth V_t and the binding energy ε, equation **11.11** serves to fix the well radius d in terms of the well depth V_t, or V_t in terms of d, for a given binding energy.

181 **The ground state of the deuteron**

11.3.3 *Solution in first approximation*

Equation **11.11**, rewritten as

$$\cot Kd = -\frac{\alpha}{K},\qquad\qquad\qquad\text{11.12}$$

shows that Kd is just over $\frac{1}{2}\pi$, since α/K is small for a deep well. As a first approximation, we may take a well depth V_t so great in comparison with ε that α/K may be neglected, and Kd put equal to $\frac{1}{2}\pi$. To this level of approximation, we may put

$$K = \frac{\sqrt{(2mV_t)}}{\hbar},$$

giving

$$d = \frac{\pi}{2K} = \frac{\pi\hbar}{2\sqrt{(2mV_t)}},$$

whence $d^2 V_t = \dfrac{\pi^2\hbar^2}{8m} = 1\!\cdot\!02 \times 10^{-24}\ \mathrm{cm^2\ MeV}$.

This argument shows that the existence of a bound state requires the well to have $d^2 V_t$ close to $10^{-24}\ \mathrm{cm^2\ MeV}$ irrespective of the binding energy, provided this is much less than the well depth. Neither the depth nor the radius of the well is fixed, only the product depth × radius². With an element of arbitrariness and the benefit of hindsight, we shall adopt a value of 40 MeV for the depth, which requires d to be 1·6 fm in this first approximation.

11.3.4 *Solution in second approximation*

A rather less crude approximation to the solution of equation **11.11** puts

$$-\cot Kd = \tan\!\left(Kd - \frac{\pi}{2}\right) = \frac{\alpha}{K}$$

and instead of ignoring the small angle $Kd - \frac{1}{2}\pi$ puts it equal to its own tangent, giving

$$Kd = \frac{\pi}{2} + \frac{\alpha}{K},$$

whence $d = \dfrac{\pi}{2K} + \dfrac{\alpha}{K^2}.$ $\qquad\qquad$ **11.13**

For $V_t = 40$ MeV and $\varepsilon = 2\!\cdot\!225$ MeV, equations **11.7** and **11.8** give

$K^{-1} = 1\!\cdot\!05$ fm,

$\alpha^{-1} = 4\!\cdot\!30$ fm.

With these values, the main term in equation **11.13** is

$$\frac{\pi}{2K} = 1\!\cdot\!64\ \mathrm{fm},$$

compared with 1·6 fm from the first-approximation calculation which used an approximate value for K.

The second term, with the same well depth, is

$$\frac{\alpha}{K^2} = 0·25 \text{ fm},$$

giving a total

$$d = 1·89 \text{ fm}.$$

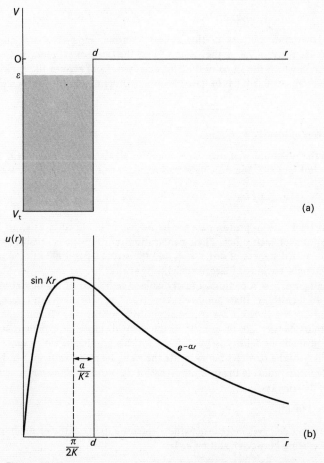

Figure 45 Neutron–proton interaction in triplet state. (a) Potential well, showing binding energy of deuteron = ε. (b) $u(r)$ = modified wave function

11.3.5 Exact solution

If we are contenting ourselves with finding the value of well radius for a given depth, we may rewrite equation **11.12** as an explicit expression for d:

$$d = K^{-1} \tan^{-1}\left[-\frac{K}{\alpha} \right].$$

This gives a numerical result confirming the second approximation, to three significant figures. We may therefore quote for the parameters of a square well satisfying the conditions deduced from the properties of the deuteron

$$d = 1.89 \,\text{fm}; \qquad V_t = 40 \,\text{MeV}. \hspace{4em} \textbf{11.14}$$

The two contributions in the second approximation to d are shown in Figure 45, where $\pi/2K$ is the value of r at the peak of $u(r)$, and α/K^2 is the distance in which the slope of the plot decreases from zero at the peak to the negative value which it has to have at $r = d$ in order to join the solution outside the well.

11.3.6 The radius of the deuteron

Inside the potential well, the wave function $u(r)$ is of the form $B \sin Kr$, as shown in Figure 45. But outside it falls off exponentially, with decay length

$$\alpha^{-1} = \frac{\hbar}{\sqrt{(2m\varepsilon)}} = 4.3 \,\text{fm}.$$

Thus the neutron–proton pair are by no means confined to existing within a distance d of each other. Their penetration into the space outside the well extends over distances of order α^{-1}, and the structure is loose in comparison with the tight packing of larger nuclei.

Although α^{-1} is sometimes loosely called the 'radius of the deuteron', it is really no such thing. Blatt and Weisskopf (1952, p. 506) point out that it is the neutron–proton distance for exponential decay of the wave function, but the probability density falls off as $u(r)^2$, so the decay distance for exponential decay of the probability density is $\frac{1}{2}\alpha^{-1}$. Thus on this argument, which ignores the possibility of the particles being inside the well, the mean value of r is $2.15 \,\text{fm}$, and the mean radius in the ordinary sense of the word (half the mean neutron–proton distance) is

$$\tfrac{1}{4}\alpha^{-1} = 1.08 \,\text{fm}.$$

This point deserves more exact study: the probability of the neutron–proton distance being between r and $r+dr$ is

$$|\psi|^2 dv = 4\pi r^2 \, dr \,|\psi|^2$$
$$= 4\pi u^2 \, dr.$$

If we take the integral of $u^2 \, dr$ for values of r from 0 to d, and then from d to ∞, we find that for the well specified by the values **11.14** there is a 34 per cent chance of the particles being inside the well and 66 per cent chance of their being outside. The mean value of r, calculated from

$$\frac{\int\limits_{0}^{\infty} r u^2 \, dr}{\int\limits_{0}^{\infty} u^2 \, dr},$$

turns out to be a rather complicated function of d, k and α, with numerical value

$$\langle r \rangle = 3 \cdot 10 \text{ fm}.$$

For convenience, some of the distances used in this section are listed in Table 13.

Table 13 Table of Dimensions of Deuteron, for Well Depth 40 MeV (All Measured across the Origin, in Units of fm $= 10^{-15}$ m)

Decay distance of wave function outside well	$\alpha^{-1} = 4 \cdot 30$
Decay distance of density outside well	$\frac{1}{2}\alpha^{-1} = 2 \cdot 15$
λ inside well	$K^{-1} = 1 \cdot 05$
Radius of well	$d = 1 \cdot 89$
First approximation to d	$\dfrac{\pi}{2K} = 1 \cdot 64$
Second term in d	$\dfrac{\alpha}{K^2} = 0 \cdot 25$
Mean neutron–proton distance	$\langle r \rangle = 3 \cdot 10$

11.3.7 *Conclusion*

The above simple treatment of the ground state of the deuteron has taken no account of the non-central forces which cause the deuteron to have an admixture of a $^3D_1$ state, leading to the existence of an electric quadrupole moment (see section 12.1). It does, however, give a substantially correct description of the main nucleon–nucleon forces in the $^3S_1$ state of total spin one. The fact that nucleon–nucleon forces in the other spin-state are different is discussed in the following section.

185 The ground state of the deuteron

11.4 Neutron–proton scattering

11.4.1 *Low-energy theory*

Having considered the neutron-proton interaction in the bound state known as the deuteron, we may use the methods of Chapter 5 to extend our studies into the region of positive energies, where the interaction leads to scattering. For simplicity we shall limit the energies to values less than a few million electronvolts, which are small enough to ensure that only S-wave scattering occurs (see section 5.3).

The potential well will be taken to be square, with depth V_t and radius d, as in the calculations on the deuteron. Thus the discussion applies only to scattering in the triplet spin-state. When we consider a scattering process in which neutrons of kinetic energy E_{inc} are incident upon stationary protons, the energy E used in the Schrödinger equation should be the total incident kinetic energy in the centre-of-mass system, or relative energy. As is shown in equation **3.18**, E is given by

$$E = \frac{m_p}{m_p + m_n} E_{inc} \simeq \tfrac{1}{2} E_{inc}.$$

Incident and elastically scattered particles should therefore be described in terms of waves with a wave number

$$k = \frac{p}{\hbar} = \frac{\sqrt{(2mE)}}{\hbar} \simeq \tfrac{1}{2} k_\infty,$$

where m is the reduced mass, p the c.m.s. momentum (see equations **3.13–17**), and k_∞ is the wave number which the same incident neutrons would have for scattering by an infinitely heavy target.

The wave function describing the state of affairs outside the potential well will be the total wave function of equation **5.2**, which for $l = 0$ and $\eta = 1$ (purely elastic scattering in S-wave only) simplifies to

$$\psi = \frac{A}{2ikr} (e^{2i\delta} e^{ikr} - e^{-ikr})$$

$$= \frac{A}{kr} e^{i\delta} \sin(kr + \delta).$$

The modified wave function $u(r)$ outside the potential well is therefore

$$u(r) = r \, \psi(r) = A k^{-1} e^{i\delta} \sin(kr + \delta). \qquad \textbf{11.15}$$

Inside the potential well, $u(r)$ will be basically the same as in the case of the deuteron, and we write it as

$$u(r) = B \sin K_0 r,$$

where $\quad K_0^2 = \dfrac{2m(V_t + E)}{\hbar^2}. \qquad \textbf{11.16}$

K_0 is not very different from the K of equation **11.7**, since the difference between the negative energy $-\varepsilon$ and the small positive energy E is much smaller than the well depth V_t.

Continuity of the wave function and its gradient across the boundary at $r = d$ requires

$$B \sin K_0 d = Ak^{-1}e^{i\delta} \sin(kd+\delta)$$

and $\quad K_0 B \cos K_0 d = Ae^{i\delta} \cos(kd+\delta),$

whence, by dividing to remove the normalizing constants A and B,

$$K_0 \cot K_0 d = k \cot(kd+\delta). \tag{11.17}$$

For sufficiently small values of E, $K_0 \cot K_0 d$ is a constant of the potential well. We call it $-\alpha_0$, and note that in the approximation that neglects the small difference between K_0 and K we have

$$-\alpha_0 = K_0 \cot K_0 d \simeq K \cot Kd,$$

whence, by equation **11.11**, α_0 is nearly equal to α, the decay distance of the wave function outside the deuteron.

Therefore, without making any approximation, but still limiting ourselves to small values of E, we may rewrite equation **11.17** in the form

$$k \cot(kd+\delta) = -\alpha_0. \tag{11.18}$$

Putting this conclusion into words, we may say that the wave function outside the well in neutron–proton scattering has to fit a wave function inside which is practically the same as the one that fits the exponentially decaying wave function outside the deuteron. Thus the function $k \cot(kd+\delta)$ which is relevant to scattering, has a value $(-\alpha_0)$ fixed by the energy of the only bound state accessible to the system.

Equation **11.18** may be rewritten

$$\tan(kd+\delta) = -\frac{k}{\alpha_0}.$$

If we go to lower and lower energies, k approaches zero and the tangent approaches the angle $kd+\delta$, so we have in the limit of small k,

$$kd+\delta = -\frac{k}{\alpha_0}.$$

In this limit the phase shift δ becomes

$$\delta = -k(\alpha_0^{-1}+d). \tag{11.19}$$

11.4.2   *Scattering length*

It thus appears that δ must approach zero as k becomes small; but their negative ratio $-\delta/k$, which is a length, tends towards a fixed value, $\alpha_0^{-1}+d$. This limiting value of $-\delta/k$ is given the special name of scattering length. We shall use the symbol a for scattering length in general, with suffix t in the present case to indicate that it refers to scattering in the triplet state of total spin 1. Thus we have shown that

$$a_t = \alpha_0^{-1} + d. \qquad \textbf{11.20}$$

The scattering length has an important significance which is shown rather directly by equation **5.13** for the differential cross-section for pure S-wave scattering. This is given as

$$\frac{d\sigma}{d\Omega} = (\lambda \sin \delta)^2, \qquad [\textbf{5.13}]$$

which, as $\delta \to 0$ and $\lambda^{-1} = k \to 0$ becomes

$$\frac{d\sigma}{d\Omega} = a^2,$$

where $a = \lim \lambda \delta$ is the scattering length. The cross-section itself, for S-wave scattering, is

$$\sigma = 4\pi \frac{d\sigma}{d\Omega} = 4\pi\lambda^2 \sin^2\delta \qquad [\textbf{5.14}]$$

$$= 4\pi a^2. \qquad \textbf{11.21}$$

Thus the scattering length for any scattering process is fixed, in magnitude though not in sign, by the value of the cross-section at very low energy.

11.4.3 *Effective-range approximation*

When the incident energy is not small enough to make equation **11.19** a good solution to equation **11.17**, it is necessary to improve upon the device of putting δ/k equal to its limiting value.

A solution which does not depend upon the shape of potential well assumed is given by Blatt and Weisskopf (1952, p. 62). (See also Segré, 1964, Appendix E, but note misprints of r_0^2 for r_0.)

An alternative general method runs as follows. The scattering amplitude for purely elastic S-wave scattering is given by equations **5.6** and **5.12** as

$$f(\theta) = \frac{\lambda e^{i\delta} \sin \delta}{2i},$$

which may be rewritten as

$$f(\theta) = \frac{\sin \delta}{2ike^{-i\delta}}$$

$$= \frac{1}{2ik(k \cot \delta - ik)}.$$

This shows that all information about the scattering is contained in the term $k \cot \delta$. The effective-range method is to approximate this term by a power series in the energy (i.e. in k^2) by putting

$$k \cot \delta = -\frac{1}{a} + \tfrac{1}{2}k^2 r_0 + \ldots, \qquad \qquad 11.22$$

where a is the scattering length, equal to the limiting value of $-\delta/k$ as in our simple treatment above. For a square well, r_0 is a little smaller than the well radius d, the difference depending on the well depth.

This means that as k increases from zero, the phase shift δ increases in a way given by

$$\frac{\delta}{k} = -a(1 + \tfrac{1}{2}k^2 a r_0).$$

The cross-section, obtained from equations **5.13** and **11.22**, reduces to

$$\sigma = \frac{4a^2}{1 + k^2 a(a - r_0) + \ldots},$$

which shows the appropriate decrease as k increases. Of course the approximation that k is small, which allowed neglect of the higher powers of k, prevents this from covering more than the initial decrease.

It is interesting to contrast the effective-range approximation with the Born approximation, a widely used alternative device for describing scattering processes. In the former the scattering can cause a severe modification to the incident waves, and we have only assumed that the energy is low. On the other hand the Born approximation assumes that scattering involves only a small perturbation of the incident waves, an assumption which tends to gain validity with increasing energy.

11.4.4 Cross-sections

From the properties of the deuteron we calculated a value of $4 \cdot 3$ fm for α^{-1}, and for a well depth of 40 MeV we obtained a value of $1 \cdot 89$ fm for the well radius d. With $E = 0$ in equation **11.16**, we obtain

$$K_0^{-1} = 1 \cdot 015 \text{ fm},$$

whence $\quad \alpha_0^{-1} = (K_0 \cot K_0 d)^{-1} = 3 \cdot 36 \text{ fm},$

which gives a value of

$$a_t = \alpha_0^{-1} + d = 5 \cdot 25 \text{ fm} \qquad \qquad 11.23$$

for the neutron–proton scattering length in the triplet state.

This would lead us to expect a cross-section

$$\sigma_t = 4a_t^2 = 3 \cdot 46 \text{ barns} \quad (1 \text{ barn} = 10^{-24} \text{ cm}^2 = 100 \text{ fm}^2).$$

But experimental measurement of the cross-section for neutron–proton scattering gives results which increase with decreasing energy, reaching 20

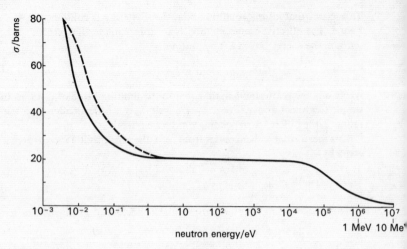

Figure 46 Cross-section per proton for neutron–proton scattering in hydrogen gas (full line) and in paraffin (broken line)

barns at about a thousand electronvolts, then remaining constant down to one electronvolt, and then increasing further to reach 78 barns at 0·003 eV. Results obtained by Melkonian (1949) with hydrogen gas are shown in Figure 46, along with those of Rainwater *et al.* (1948) for protons in paraffin.

The rise at energies below one electronvolt is due, in the case of paraffin, to the fact that the protons are bound chemically to a solid lattice. When the neutrons become too slow to dislodge them, the scattering follows kinematics appropriate to an almost infinitely heavy target, which give a cross-section four times that for free protons. In the case of hydrogen gas, the increase at lowest energies is due partly to this increase of effective mass (which gives a factor $\frac{16}{9}$ for an H_2 molecule, cf. 4 for a heavy lattice), and partly to coherent scattering by the two protons of each hydrogen molecule when the wavelength of the neutrons is no longer small compared with the distance between the protons. With the rise at lowest energies explained in this way, all the experimental information points towards a value close to 20·4 barns as the low-energy cross-section which must be fitted to the theory.

By choosing different well parameters it is possible to obtain some spread in the calculated values of $4\pi a_t^2$, but there is no possibility of explaining more than about a quarter of the experimental value of 20·4 barns.

From this we conclude that the neutron–proton interaction is spin dependent, and that the scattering cannot be treated as if it all occurred in the triplet state with neutron and proton spins parallel.

11.4.5 *Scattering in the singlet state*

A neutron and a proton have four spin-states, one singlet and three components of the triplet. If all the particles are unpolarized, these four states should be equally probable, so three-quarters of neutron–proton interactions should occur in triplet states, and a quarter in singlet states. If the cross-sections for scattering in the two types of state are σ_t and σ_s, the mean cross-section, per proton, should be

$$\sigma_{\text{mean}} = \tfrac{3}{4}\sigma_t + \tfrac{1}{4}\sigma_s. \qquad \textbf{11.24}$$

Putting in the observed value of 20·4 barns for σ_{mean}, and the calculated value of 3·46 barns for σ_t, we get for the cross-section for scattering in the singlet state a value of

$$\sigma_s = 71\cdot2 \text{ barns}.$$

This surprisingly high cross-section suggests a singlet scattering length a, of magnitude

$$|a_s| = \sqrt{\frac{\sigma_s}{4\pi}} = 23\cdot8 \text{ fm}.$$

Thus if the singlet state were a bound state, with large well depth and small well diameter, it would have a decay distance outside the well equal to

$$\alpha_s^{-1} \simeq 23\cdot8 \text{ fm},$$

and therefore a binding energy ε_s given by

$$\alpha_s = h^{-1}\sqrt{(2m\varepsilon_s)}.$$

This may be calculated, in terms of the parameters of the triplet state, as

$$\varepsilon_s = \left(\frac{\alpha_s}{\alpha_t}\right)^2 \varepsilon_t = 72 \text{ keV}.$$

No such bound state has been observed, so we conclude that the singlet state is a virtual one, with an interaction such that it creates as much scattering as a bound state at 72 keV would have done. For such a virtual state, the plot of u against r for small k is as shown in Figure 47(b).

Figure 47 is drawn to illustrate a further significance of the scattering length a. Since a is defined as the value of $-\delta/k$ for small k, $\sin(kr+\delta)$ is zero where $r = a$. Therefore if r becomes equal to a somewhere outside the potential well, where $u(r)$ is given by equation **11.15**, $u(r)$ must be zero there. This happens for an attractive potential with a bound state, and Figure 47(a) shows the positive triplet scattering length as the value of r at which the plot of $u(r)$ crosses the axis.

But in the singlet case, shown in Figure 47(b), $u(r)$ does not itself cross the axis; $\sin(kr+\delta)$ is zero at the intercept obtained by extrapolating the straight plot of $u(r)$ back to the r-axis. This intercept represents the singlet scattering length a_s, which must therefore be negative. Thus we can distinguish whether

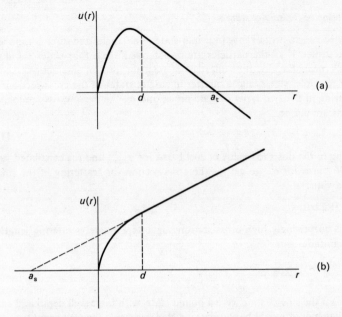

Figure 47 Modified wave function $u(r)$ for neutron–proton scattering at small values of k, showing scattering lengths a_s and a_t. (a) Interaction in triplet state, scattering length positive. (b) Interaction in singlet state, no bound states, scattering length negative

or not a potential gives rise to a bound state by the sign of the scattering length, and specify the unbound nature of the singlet neutron–proton interaction by writing

$$a_s = -23\cdot8 \text{ fm}. \qquad \textbf{11.25}$$

If the singlet interaction is to be described in terms of a square well, values for the parameters of the well (depth V_s and radius d_s) may be obtained from the above value of a_s by arguments similar to those used in sections 11.4.1 and 11.4.2 for scattering in the triplet state. The equivalent of equation **11.20** may be written

$$a_s = \alpha_{0s}^{-1} + d_s,$$

where α_{0s} is given (see equation **11.17** *et seq.*) by

$$\alpha_{0s} = K_s \cot K_s d_s.$$

K_s is the wave number of a zero-energy neutron inside the well, which is related to the well depth V_s by

$$K_s = \frac{\sqrt{(2mV_s)}}{\hbar}.$$

Inserting the best value of a_s ($-23{\cdot}7$ fm, see equation **11.28**) fixes the value of V_s for a given d_s, typical values being

$d_s/$fm	$V_s/$MeV
2·1	22
2·5	15
2·8	12

To choose a best pair of values from this range, we need further information (see section 11.4.7).

11.4.6   *Coherent scattering*

As has been mentioned already, neutrons of thermal energy have wavelengths which may be of the same order as the distances between nuclei of adjacent atoms in a molecule. In this case, the total scattering is obtained, not by adding the intensities or cross-sections for scattering by the individual nuclei, but by adding the amplitudes of the scattered waves from all the scattering nuclei and squaring the total amplitude to get the intensity.

In a regular solid, the interference of these waves may cause concentration of the scattered neutrons into a small number of well-defined directions, allowing studies of crystal structure closely analogous to those made by the methods of X-ray crystallography. In this situation, the scattered intensity is not, in general, proportional to the number of target nuclei. The concept of cross-section therefore has to be used with care; special methods are needed for describing the cooperative effects resulting from the nature of the lattice of scattering centres.

Between this extreme case and that of the single, independent, scattering nucleus, there comes scattering by molecules. Especially interesting in this category is the scattering of slow neutrons by molecules of gaseous hydrogen. As will be explained in section 12.2.1, hydrogen molecules can exist in either of two states, namely orthohydrogen with the two protons in a triplet state of total spin 1, and parahydrogen with the two protons in a singlet state of total spin zero.

If the two protons in a hydrogen molecule are causing coherent scattering of a beam of neutrons, the amplitudes of the wave functions representing scattering of a neutron by each of the two protons must be added. This may be done by adding the scattering lengths for the two n + p interactions, putting each equal to a_t if the neutron and proton are in a triplet state, or to a_s if they are in a singlet state. The correlation between the two n + p states depends on the correlation of the two proton spins, so triplet and singlet scattering contribute to different extents in the two types of hydrogen molecule. The formal treatment of this problem, using projection operators to pick out singlet and triplet contributions, with correct amplitudes and relative phases, will be found in the literature (e.g. Blatt and Weisskopf, 1952, p. 82; Evans, 1955, p. 324; Segré, 1964, p. 572 – but note the factors $\frac{4}{3}$ included in the scattering lengths, and the brackets omitted in σ_{ortho}). The result is a total scattering length for a para-hydrogen molecule

$$a_{\text{para}} = 2(\tfrac{3}{4}a_t + \tfrac{1}{4}a_s).$$

The extra effective mass, mentioned in section 11.4.4, gives a factor $\tfrac{16}{9}$ in the cross-section for scattering by hydrogen molecules, so the cross-section for scattering of slow neutrons by parahydrogen, per molecule, is

$$\sigma_{\text{para}} = \tfrac{16}{9}\pi(3a_t + a_s)^2.$$

The corresponding expression for orthohydrogen contains two incoherent terms, giving a total cross-section per molecule

$$\sigma_{\text{ortho}} = \tfrac{16}{9}\pi\{(3a_t + a_s)^2 + 2(a_t - a_s)^2\}.$$

If we substitute in these expressions the values for a_t and a_s given in equations **11.23** and **11.25**, we get predicted cross-sections per molecule,

$\sigma_{\text{para}} = 3{\cdot}6$ barns,

$\sigma_{\text{ortho}} = 97$ barns.

For comparison of these predicted values with observation, we refer to the experiments of Sutton *et al.* (1947). In order to obtain the best possible approximation to the ideal situation of zero-energy neutrons being scattered by stationary hydrogen molecules, they used thermal neutrons separated by velocity selection to give beams of energy from $0{\cdot}0025$ eV down to $0{\cdot}0008$ eV, equivalent to thermal energies at temperatures from 30 K to 10 K. The targets were gaseous hydrogen at 20 K, either parahydrogen (see section 12.1.4), or the normal mixture of three-quarters orthohydrogen and a quarter parahydrogen. At their lowest energy, their observations indicated cross-sections

$\sigma_{\text{para}} = 4{\cdot}98$ barns,

$\sigma_{\text{ortho}} = 145$ barns.

They point out that these figures still contain factors resulting from the thermal motion of the target molecules, and in the case of orthohydrogen from induced transitions to the para state. Figures for ideal conditions, extracted from these raw data, are

$\sigma_{\text{para}} = 3{\cdot}6$ barns,

$\sigma_{\text{ortho}} = 92$ barns.

The agreement with prediction is satisfactory in both cases, for the measurements over the whole energy range indicate that σ_{para} will change little between $0{\cdot}0008$ eV and zero energy, while σ_{ortho} could well rise five barns.

Less satisfaction is obtained from Melkonian's figure of 80 barns per proton (160 per molecule), obtained with $0{\cdot}003$ eV neutrons in hydrogen gas at room temperature (see Figure 46). He estimates that allowance for thermal motion should bring this figure down to 55 barns per proton, but this is still too high, and we may infer from the work at lower target temperature and neutron energies that an even more drastic correction would be needed.

11.4.7 *Best values*

The calculations presented above have all been dependent upon the value of the triplet scattering length calculated from a square potential well, which is not particularly realistic. It gives better agreement than might have been expected, in the values of the low energy cross-sections for scattering of neutrons by orthohydrogen and parahydrogen. Ultimately the sequence of argument must be reversed, the two scattering lengths being deduced from experiment, and the parameters of the well from the scattering lengths.

The singlet and triplet scattering lengths may be obtained by simultaneous solution of the equations for two measurable quantities, namely:

$$\sigma_{\text{mean}} = 4\pi(\tfrac{3}{4}a_t^2 + \tfrac{1}{4}a_s^2), \qquad\qquad\qquad \textbf{[11.24]}$$

$$a_{\text{coh}} = \tfrac{3}{4}a_t + \tfrac{1}{4}a_s. \qquad\qquad\qquad \textbf{11.26}$$

The latter quantity, known as the coherent scattering length, is the average scattering length for coherent scattering of slow neutrons by randomly oriented protons. It is the same as the scattering length per proton in parahydrogen, which behaves for this purpose as an assembly of randomly oriented protons. The value of a_{coh} may be obtained from the measured value of σ_{para}, but an equivalent and possibly more accurate value is given by an experiment of Burgy, Ringo and Hughes (1951). This was a direct measurement of a_{coh} by reflection of slow neutrons from the surface of liquid hydrocarbons, giving a value of

$$a_{\text{coh}} = -(1\cdot89 \pm 0\cdot01)\,\text{fm}.$$

For σ_{mean}, which is the mean cross-section per proton for low-energy incoherent scattering, we use Melkonian's value,

$$\sigma_{\text{mean}} = 20\cdot36 \pm 0\cdot10\ \text{barns}.$$

Insertion of these values in equations **11.24** and **11.26** gives the following best values:

$$a_t = 5\cdot38\ \text{fm}, \qquad\qquad\qquad\qquad\qquad \textbf{11.27}$$

$$a_s = -23\cdot7\ \text{fm}. \qquad\qquad\qquad\qquad\qquad \textbf{11.28}$$

A triplet square well of depth 40 MeV, with radius fixed by the binding energy of the deuteron, gave a value of $5\cdot25$ fm for a_t. Repeating the calculations of sections 11.3.5 and 11.4.4 for other well depths, we find that the above best value of a_t is given by a square well of depth $V_t = 33$ MeV, for which the binding energy of the deuteron fixes the radius as $d = 2\cdot13$ fm. These may therefore be taken as the best parameters to use when the triplet neutron–proton interaction is to be described by a square well.

The best available information about the singlet interaction is contained in the value of its scattering length ($a_s = -23\cdot7$ fm) which, as we have seen in section 11.4.5, relates the depth to the radius for a potential well of given shape. To determine the actual radius of the singlet potential well, we use the energy dependence of low-energy singlet scattering, which may be described (see section 11.4.3) in terms of an effective range of $2\cdot4$ fm. This value allows us to choose, from the results of section 11.4.5, best values:

$$d_s \simeq 2\cdot5 \text{ fm}, \qquad \qquad \textbf{11.29}$$

$$V_s \simeq 15 \text{ MeV}. \qquad \qquad \textbf{11.30}$$

In fact there is considerable spread in the values quoted for these quantities, and it should be remembered that square wells are intended to provide easily handled approximations rather than detailed description. More advanced work is needed to develop a shape-independent theory which covers the different types of well in a general way. Such a description includes an effective range (see section 11.4.3) related to the energy dependence of the scattering cross-section. A given effective range implies a different radius for different assumptions about the shape of the well. In general, the scattering length and the effective range provide the most direct specification of the interaction.

Further reading

H. A. Bethe and P. Morrison, *Elementary Nuclear Theory*, 1956, Wiley.

W. E. Burcham, *Nuclear Physics*, 1963, Longman.

H. A. Enge, *Introduction to Nuclear Physics*, 1966, Addison-Wesley.

M. A. Preston, *Physics of the Nucleus*, 1962, Addison-Wesley. See also Blatt and Weisskopf (1952) and Segré (1964).

Chapter 12
Nuclear Spin

12.1 Spin and magnetic moment

12.1.1 *The nature of nuclear spin*

A nucleus may have angular momentum resulting from the spins of the nucleons from which it is made up, and also from any orbital angular momentum of the relative motion of the nucleons. When thinking of the structure of the nucleus in terms of the shell model, we may want to distinguish between these contributions, using the symbol $\mathbf{S}$ for the resultant of the spins of the individual nucleons, and $\mathbf{L}$ for the resultant of the orbital angular momenta of the states of motion occupied by the nucleons. The total angular momentum of the nucleus is then the vector sum

$$\mathbf{I} = \mathbf{L} + \mathbf{S}.$$

However, when one is considering the effect of the nucleus as a whole, in determining, for example, the hyperfine structure of atomic spectra or the energy resulting from orientation of a nucleus in a magnetic field, what matters is $\mathbf{I}$ itself, not the structure that gives rise to it.

$\mathbf{I}$ may then be combined with the spin of an atomic electron, and any orbital angular momentum of the electron orbit, to give the total angular momentum of the atom. So the term 'total angular momentum' means one thing to a shell-model man and another to a spectroscopist. To avoid having to use the clumsy term 'total nuclear angular momentum' for $\mathbf{I}$ we just call it the nuclear spin. This is a convenient and unambiguous name for the observable quantity.

12.1.2 *Values of nuclear spin*

The mass number A of a nucleus represents the total number of nucleons in it, and each nucleon has spin one half. Pairs of nucleons can have total spin 0 or 1, and orbital angular momenta are integral multiples of the same unit $\hbar$, so, however these combine, a nucleus of even A must have integral (including zero) nuclear spin I. Correspondingly, all nuclei with odd A have half-integral nuclear spin I.

This fact, which seems obvious nowadays, was once crucial as evidence that nuclei contained neutrons and not electrons. For example Rasetti (1929) pointed out the discrepancy between the nuclear spin $I = 1$ which he observed for nitrogen-14 and the nuclear constitution which was at that time assumed for nitrogen-14 (fourteen protons and seven electrons).

In fact, stable nuclei with even A all have $I = 0$ in the ground state, except for the small odd–odd nuclei (i.e. odd Z, odd N, making even A) ^2_1H, ^6_3Li, $^{10}_5B$ and $^{14}_7N$.

12.1.3 *The Bohr magneton*

If the electron followed perfectly the simplest form of the Dirac model for a point charge, it would have magnetic moment equal to one Bohr magneton, a unit given by

$$\mu_B = \frac{e\hbar}{2m_e c},\qquad\qquad\textbf{12.1}$$

where m_e is the mass of the electron.

But in fact there are corrections due to radiative effects and vacuum polarization. These have been calculated on the basis of quantum electrodynamics, and indeed it is one of the triumphs of this subject that the calculated value of the correction (1159·615 parts per million) agrees so well with the experimental value

$$(1159\cdot622 \pm 0\cdot027) \times 10^{-6}$$

obtained by Wilkinson and Crane (1963).

However, the Bohr magneton remains the natural unit of magnetic moment for particles of mass m_e. The correction merely means that the electron itself has a magnetic moment

1·001 159 622 ± 0·000 000 027   Bohr magnetons.

For other types of particle, the natural unit of magnetic moment is a magneton obtained by inserting the appropriate mass instead of m_e in equation **12.1**.

12.1.4 *The nuclear magneton*

For a proton, the natural unit of magnetic moment is the proton magneton

$$\mu_N = \frac{e\hbar}{2m_p c},\qquad\qquad\textbf{12.2}$$

which is smaller than the Bohr magneton by a factor

$$\frac{m_e}{m_p} = \frac{1}{1836}.$$

The proton is much farther than the electron from being a perfect Dirac particle, for the corrections amount to a factor greater than unity, and the observed magnetic moment of the proton is

$\mu_p = 2 \cdot 793$ proton magnetons.

Although $e\hbar/2m_p c$ is properly the proton magneton, it has come to be known as the nuclear magneton and to be used as the general unit for all nuclear moments. This is the reason for the suffix N in the symbol μ_N of equation **12.2**.

12.1.5   *Magnetic moment of the neutron*

If the neutron were a neutral particle in the sense of having no electromagnetic properties, it would have no magnetic moment. But in fact it is found to have a magnetic moment

$\mu_{\text{neutron}} = -1 \cdot 913$ nuclear magnetons.

12.1.6   *Sign of magnetic moment*

Normally the term 'magnetic moment' is used either for the vector quantity whose scalar product with a magnetic-field vector gives the energy of interaction with the field, or for the magnitude of this vector quantity; but in the above we have been quoting magnetic moments of particles as scalar quantities with signs. This is because the magnetic moment of a particle arises from the rotation of the charge carried by it. Thus the magnetic-moment vector must be directed along the axis of spin, in one sense or the other. The quantity quoted is the magnetic moment measured in the direction of the spin vector. The vectors of magnetic moment and spin are both axial vectors, that is, they are defined by vector products which can be in one sense or the other according to whether the sign convention adopted uses a left-handed or a right-handed set of axes. Thus the same element of convention affects the two quantities, but the definition of magnetic moment involves also a choice of sign of charge or current. In fact this choice is made so that a rotating positive charge has magnetic-moment vector in the same sense as its spin vector.

Thus when we quote a negative value for the magnetic moment of the neutron, we are saying that its magnetic moment is directed in the sense opposite to the spin, that is, that its magnetic effects are like those of a spinning negative charge. But it is electrically neutral, so if we imagine it as a sphere rotating with uniform angular velocity, we must give it a non-uniform charge distribution, with negative charge outside and positive inside.

It should be noted that the electron is a spinning negative charge, and has a magnetic-moment vector directed oppositely to its spin vector. Therefore the above convention, applied consistently, should make us quote the magnetic moment of the electron as

$-1 \cdot 001\ 159\ 622$ Bohr magnetons.

This point is important in relation to the magnetic splitting of atomic levels.

12.1.7 Nuclear magnetic moments

It has already been mentioned (section 11.3) that the deuteron has magnetic moment 0·857 nuclear magnetons, which fits quite accurately the total magnetic moment expected for a neutron and a proton, lined up with spins parallel.

But what happens when a neutron and a proton have their spins antiparallel? One might think that if they were aligned in opposite directions there would be a total magnetic moment of 2·793 + 1·913 nuclear magnetons. However, this situation is never observed, for a system of neutron and proton with spins antiparallel has zero angular momentum and is without any preferred direction. Thus if we try to measure the magnetic moment as the slope of a plot of energy against magnetic field, we must get a result zero. The alternative method of obtaining magnetic moment from the energy difference per unit magnetic field between two states with different components of angular momentum along the direction of the field, is ineffective when the total angular momentum is zero.

Thus nuclei with even A and zero nuclear spin I have zero magnetic moment. Nuclei with non-zero I have magnetic moments which may be positive or negative. In the deuteron, the magnetic moment corresponds numerically with that expected from the unbalanced spins of the component nucleons. In larger nuclei the correspondence is more qualitative than quantitative. For example, for nuclei in their ground states, the largest listed value of I is seven, for the β-active nucleus $^{176}_{71}$Lu, with magnetic moment 3·18 nuclear magnetons, and the largest listed magnetic moment is 6·167 nuclear magnetons for $^{93}_{41}$Nb, with nuclear spin $I = \frac{9}{2}$.

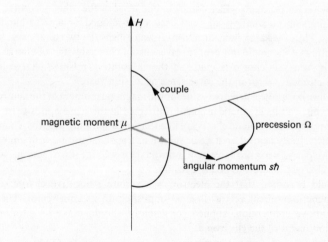

Figure 48 Precession due to couple acting on a magnetic moment μ in a magnetic field H. The particle has an angular momentum $s\hbar$ along the same axis as its magnetic moment

Measurement of magnetic moments

Most nuclear magnetic moments have been measured by atomic beam methods, which are discussed briefly in section 12.2.6. The precision quoted for results of this type is usually of order 10^{-4} magneton. The reader may therefore wonder at the high accuracy (2×10^{-8} magnetons) of the value quoted for the electron in section 12.1.3. The explanation lies in the fact that a special type of measurement is possible for a particle with magnetic moment close to one magneton (its own magneton, with its own mass inserted in $e\hbar/2mc$).

The angular velocity of precession (see Figure 48) of a particle of spin s (in units of $\hbar$) and magnetic moment μ (in units of $e\hbar/2mc$) is given by the elementary expression

$$\text{Angular velocity of precession} = \frac{\text{couple acting}}{\text{angular momentum}},$$

which becomes in this case

$$\Omega = \frac{\mu(e\hbar/2mc)H}{s\hbar}$$

$$= \frac{\mu}{s} \frac{e}{2mc} H$$

$$= g \frac{e}{2mc} H, \qquad\qquad 12.3$$

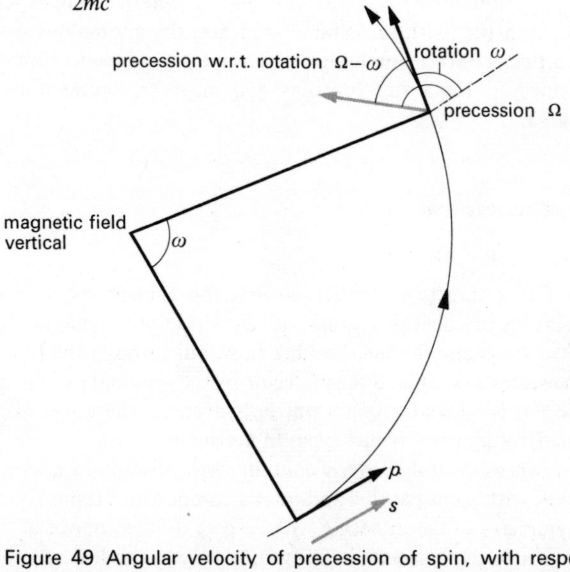

Figure 49 Angular velocity of precession of spin, with respect to rotation of the momentum vector when a charged particle with spin is moving in a circular orbit in a uniform magnetic field

201 Spin and magnetic moment

where $g = \mu/s$, which is commonly known as the gyromagnetic ratio.

If, while precession is occurring, the particle is moving through the magnetic field, it will be tracing out part of a circular orbit, its momentum vector rotating with an angular velocity

$$\omega = \frac{e}{mc} H \quad \text{(see Chapter 2).} \qquad \qquad \textbf{12.4}$$

Both these angular velocities are about the direction of the magnetic field, so the angular velocity of precession with respect to the direction of motion of the particle, as illustrated in Figure 49, is

$$\Omega - \omega = (g-2)\frac{e}{2mc} H. \qquad \qquad \textbf{12.5}$$

Thus if a polarized beam of particles which are almost perfect Dirac particles in having μ very close to one magneton (i.e. $g \simeq 2$) is trapped in a magnetic field where it can trace out repeated circular orbits, the polarization will alternate between transverse and longitudinal with a frequency proportional to $g-2$. A perfect Dirac particle would precess as fast as it rotated, and its spin direction would remain constant with respect to its direction of motion. Observing the alternations of polarization is the basis of the so-called 'g-minus-two' experiments, in which the departure from perfect Dirac behaviour has been measured for the electron (Wilkinson and Crane, 1963) and also for the muon (Farley *et al.*, 1966). The amount by which μ differs from one is some-times called the anomalous magnetic moment. A g-minus-two experiment really serves to measure, with reasonable accuracy, the anomalous magnetic moment. When this is only a small fraction of the total magnetic moment, the precision obtained in the result for the total magnetic moment is correspondingly higher.

12.2 Atomic effects of nuclear spin

12.2.1 *Orthohydrogen and parahydrogen*

In a molecule containing two identical atoms, the nuclear spins may have important effects on physical properties, not by virtue of the energy of interaction of nuclear magnetic moments with a field, but through the limitations imposed on the states available to the molecule by the symmetry of the nuclear spins. This effect is particularly important in hydrogen, where it leads to the possibility of separating two distinct types of molecule.

The type known as orthohydrogen contains two protons in a symmetric spin-state, that is, with spins parallel and total z-component of spin 1, 0 or -1. The over-all symmetry of the molecule with respect to interchange of the two protons involves also the rotational quantum number J; to have the over-all antisymmetry required by the fact that the two protons are fermions, ortho-hydrogen can exist only in rotational states with odd J.

Conversely, the parahydrogen molecule contains two protons in an anti-symmetric spin-state (a singlet state with total spin zero). Over-all antisymmetry of the molecule with respect to interchange of the two protons therefore limits parahydrogen to rotational states with even J (including 0).

Thus the two forms of hydrogen will contain different amounts of rotational energy at a given temperature, and will have different specific heats. At room temperature, the equilibrium concentration is hardly affected by the energy differences and is effectively 3:1 in favour of orthohydrogen which has three spin-states to parahydrogen's one. But at low temperatures the equilibrium ratio swings in favour of parahydrogen, for which the state of zero rotational energy is allowed. Nearly pure parahydrogen is obtained by cooling liquid hydrogen to a few kelvins, with a catalyst to increase the rate of transition to the energetically favoured para state.

The possibility of obtaining pure parahydrogen for use in neutron scattering experiments has been mentioned in Chapter 11, where it was shown that the difference between orthohydrogen and parahydrogen allowed important studies of the spin-dependence of the neutron–proton interaction.

12.2.2 *Hyperfine structure in atomic spectra*

The total energy of an atom depends on the orientation of its nuclear spin I in the magnetic field H produced at the nucleus by the electrons. The component of I in the direction of H can have any of the values I, $I-1$, ..., $-I$. The energy of interaction of the nuclear magnetic moment μ with the field H, for each of these states, is

$$\mu H, \mu H \frac{I-1}{I}, ..., -\mu H.$$

Thus a single atomic level may be split into $2I+1$ levels, and transitions leading to spectral lines may show a corresponding hyperfine structure. The magnitude of the hyperfine splitting is typically enough to give a frequency separation of order 10^9 Hz.

12.2.3 *Molecular spectra*

In the spectra of molecules, lines resulting from transitions between two electronic states are spread into bands consisting of closely spaced lines as shown in Figure 50. The spacing between the lines in a band represents the energy difference between successive rotational energy levels of the molecule. These levels will have total angular momentum J alternately odd and even. Homonuclear diatomic molecules can exist in either ortho or para form. For nuclei with half-integral spin, ortho molecules can have only rotational levels with odd J, and para molecules only levels with even J; the converse holds for nuclei with integral spin. Thus lines in a spectral band come alternately from

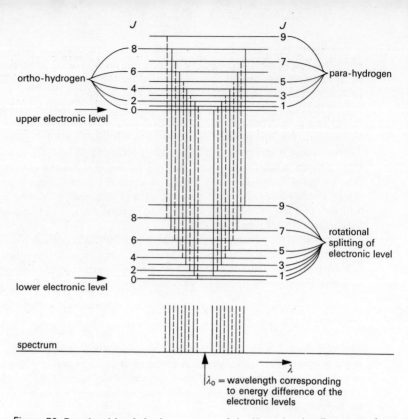

Figure 50 Rotational levels in the spectrum of the H_2 molecule: diagram to show how a single line due to transition between two electronic levels is split into a number of closely spaced lines, each corresponding to a pair of initial and final rotational states with $\Delta J = 1$. Levels with J even belong to ortho-hydrogen, and therefore give line three times more intense than those from levels with J odd, which belong to para-hydrogen. Full lines indicate transitions from ortho-hydrogen. Broken lines indicate transitions from para-hydrogen

ortho and para molecules. The two types of molecule occur in equilibrium proportions $I + 1 : I$, that is, $3 : 1$ in a case like hydrogen for which the nuclear spin I is one half. It follows that molecular band spectra show alternations of intensity which may be used to give a measure of the nuclear spin I.

12.2.4 *Microwave and radio-frequency spectra*

The hyperfine splitting, and the rotational energy of molecules, have already been mentioned as effects which can cause measurable results when superposed on an ordinary optical spectrum.

Both these effects have quantum energies which, when not superposed on the larger energy of an electronic transition, correspond to microwave frequencies. Thus microwaves of particular frequencies may be absorbed by transitions between rotational levels. Consequently molecules may be detected by observation of their characteristic microwave absorption spectra.

A well-known astrophysical phenomenon is the 21 cm line of frequency $1\cdot43 \times 10^9$ Hz, which arises from reversal of the proton spin direction in interstellar hydrogen atoms. Superposed on an optical spectral line, this transition would be called hyperfine structure.

12.2.5 *Nuclear magnetic resonance*

The 21 cm line in atomic hydrogen results from the energy difference of the two orientations of the proton in the magnetic field of the atomic electron, which is about 3×10^5 G. There exist organic compounds containing protons which are under the influence of no such electronic magnetic field. When an externally applied magnetic field H creates an energy difference $2\mu_p H$ between the two possible states of such protons, radio-frequency excitation at frequency $2\mu_p H/h$ can lead to resonant absorption as illustrated in Figure 51. Measurement of the frequency at which this resonant absorption occurs is an accurate and stable method of measuring magnetic fields.

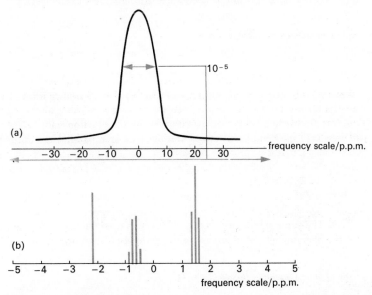

Figure 51 Proton magnetic resonance spectrum. (a) Typical low-resolution resonance plot, with peak of width 10×10^{-6}, e.g. 500 Hz in 50 MHz. (b) High-resolution spectrum showing chemical shifts due to binding of protons in ethyl alcohol. The individual lines have width of order 4×10^{-8}, e.g. 2 Hz in 50 MHz. The frequencies are plotted as differences in parts per million from the centre of the resonance, for constant magnetic field (11 743 G for resonance at 50 MHz)

12.2.6 *Atomic and molecular beam experiments*

Resonance between a radio-frequency exciting field and transitions between different substates of a nuclear magnetic moment in an externally applied magnetic field may be detected by the absorption of radio-frequency energy, as mentioned above.

A more sensitive method of detection is provided by passing the particles as a beam through two inhomogeneous magnetic fields in succession, with a uniform magnetic field and a radio-frequency exciting coil between them. The first inhomogeneous field separates one magnetic substate by a type of Stern–Gerlach experiment, and the second serves to detect the presence of particles in different substates, to which they may have been flipped by the radio-frequency excitation.

Beam methods have been developed to a high level of sophistication for specific measurements, and have given accurate values for many nuclear magnetic moments, including those of the neutron and the deuteron (see, e.g., reviews by Ramsey, 1953; Commins, 1967, and lists of results by Strominger, Hollander and Seaborg, 1958). Measurements on deuterium have provided the further information that the deuteron has an electric quadrupole moment, important evidence for non-central nuclear forces (see Chapter 11).

12.3 Spin dependence of nuclear forces

12.3.1 *Spin-orbit coupling*

The shell model of the nucleus appeared inadequate to explain the sequence of individual-particle levels until it was realized that allowance must be made for the energy of interaction between the spin of a nucleon and the orbital angular momentum *l* of the state it occupies. This interaction leads to a splitting of all except the S-states into two substates with spin parallel and antiparallel to *l*. Indeed, it is the spin–orbit splitting of the 1f, 1g and 1h levels that gives rise to the discontinuity of last-nucleon binding energy at the 'magic numbers' 28, 50 and 82 (see Reid, 1972).

It would thus be argued that the spin–orbit interaction for nucleons inside nuclei was one of the most direct manifestations of spin dependence of nuclear forces. The same interaction, for nucleons outside the nucleus, is responsible for the phenomenon of polarization by scattering (see section 12.4).

12.3.2 *Role of spin-dependent forces in nuclei*

In the building up of nuclei, a predominating factor is the complete filling of one-particle levels, each with two protons and two neutrons. One can set up a description in terms of filling the lowest level first, and then the other levels in order of increasing energy, but this fails if the levels are degenerate or very close. A stronger argument (see Blatt and Weisskopf, 1952, p. 215) goes as follows: Majorana exchange gives an attractive force between two particles with wave function symmetric with respect to space exchange, but the force is repulsive if the wave function is antisymmetric. The space wave function of a pair of nucleons is symmetric if they are in the same level, but antisymmetric or mixed if they are in different levels.† Therefore Majorana exchange gives most attraction if the nucleons are concentrated into full levels. This effect leads to the strong tendency of nuclei to contain equal numbers of neutrons and protons.

The first important modification of this tendency comes from Coulomb forces which increase the energy needed to put a proton in a given level, and in the larger nuclei lead to a situation where the stablest species have more neutrons than protons.

A further spin-dependent factor, important in nuclei containing n complete levels and two extra nucleons (i.e. $A = 4n + 2$), is the spin–spin interaction between the extra pair of nucleons. This effect provides a mechanism whereby, for example, the i-spin singlet ground state of nitrogen-14, with ordinary spin $I = 1$, is stabler than the i-spin triplet of zero ordinary spin, of which the three members are the ground states of oxygen-14 and carbon-14 and the 2·31 MeV excited state of nitrogen-14 (see section 11.3).

12.3.3 *Spin dependence of nucleon–nucleon force*

The only bound state of two nucleons is the deuteron, with i-spin $T = 0$, and ordinary nuclear spin $I = 1$. The fact that the neutron–proton interaction in the singlet spin-state is unbound (see section 11.4), while the triplet state is a bound state, constitutes a very direct indication of the interaction between the two nucleon spins.

Analysis of proton–proton scattering shows that the nuclear part of the proton–proton interaction is very similar to the neutron–proton interaction, with similar scattering lengths and spin-dependence (Breit, Condon and Present, 1936, Breit *et al.* 1960, Hull *et al.* 1961).

As we have already seen in Chapter 11, the nucleon–nucleon force can be a conservative, central force described by a potential, even though it is dependent on the spin directions of the nucleons. We simply used one potential V_t for the triplet state in which the nucleon spins are parallel, and a different potential V_s for the singlet state in which they are antiparallel. If we wish, we may replace V_s and V_t by a single potential V containing a spin-dependent term

$$V = V_d + V_\sigma(\boldsymbol{\sigma}_1 \cdot \boldsymbol{\sigma}_2).$$ **12.6**

†Within one level there is no orbital angular momentum of relative motion; the over-all antisymmetry necessary for the wave function of a pair of fermions is provided by the spin and i-spin wave functions, which allow each level to be occupied by two protons of opposite spin and two neutrons of opposite spin, in accordance with the simple form of the Pauli principle.

But two nucleons in different levels, if they happen to have spins parallel (or i-spin $T = 1$, but not both) must have orbital angular momentum odd, and space wave function antisymmetric.

$\boldsymbol{\sigma}_1 \cdot \boldsymbol{\sigma}_2$ is the scalar product of the Pauli spin operators representing the spins of the two nucleons (see, e.g., Schiff, 1968, p. 374). The product $\boldsymbol{\sigma}_1 \cdot \boldsymbol{\sigma}_2$ has eigenvalue -3 for the singlet state or $+1$ for the triplet state.† V_d is the potential averaged over all orientations of the spins, and V_σ specifies the magnitude of the spin dependence.

12.3.4 Tensor forces

So far we have considered only central forces, that is, forces directed along the line of centres of the interacting nucleons. The existence of non-central forces is demonstrated by the fact that the deuteron has an electric quadrupole moment. This is attributed to a tensor force analogous to that occurring between two bar magnets.

The tensor force is usually described by a potential of form

$$V = V_\mathrm{t} S_{12}, \tag{12.7}$$

where S_{12} is the tensor operator

$$S_{12} = \frac{3(\boldsymbol{\sigma}_1 \cdot \mathbf{r})(\boldsymbol{\sigma}_2 \cdot \mathbf{r})}{r^2} - \boldsymbol{\sigma}_1 \cdot \boldsymbol{\sigma}_2.$$

†The easiest way to visualize this is to express the total spin by an operator

$$\mathbf{S} = \tfrac{1}{2}(\boldsymbol{\sigma}_1 + \boldsymbol{\sigma}_2).$$

The individual nucleon spins are given by

$$\mathbf{s}_1 = \tfrac{1}{2}\boldsymbol{\sigma}_1 \quad \text{and} \quad \mathbf{s}_2 = \tfrac{1}{2}\boldsymbol{\sigma}_2.$$

The magnitude of the total spin is given by the operator equation

$$\mathbf{S}^2 = \tfrac{1}{4}(\sigma_1^2 + \sigma_2^2 + 2\boldsymbol{\sigma}_1 \cdot \boldsymbol{\sigma}_2).$$

$\mathbf{S}^2$ is an operator with eigenvalue $S(S+1)$, where S is the total spin, 0 for the singlet state or 1 for the triplet. Similarly σ_1^2 and σ_2^2 have eigenvalues $\tfrac{1}{2}(\tfrac{1}{2}+1)$, reflecting the fact that the Pauli matrices satisfy

$$\sigma_x^2 = \sigma_y^2 = \sigma_z^2 = 1.$$

The scalar product operator $\boldsymbol{\sigma}_1 \cdot \boldsymbol{\sigma}_2$ therefore has eigenvalue

$$\langle \boldsymbol{\sigma}_1 \cdot \boldsymbol{\sigma}_2 \rangle = 2\langle \mathbf{S}^2 \rangle - \tfrac{1}{2}(\langle \sigma_1^2 \rangle + \langle \sigma_2^2 \rangle)$$

$$= \begin{cases} 0 - 3 = -3 & \text{for the singlet state.} \\ 4 - 3 = +1 & \text{for the triplet state.} \end{cases}$$

Since $S = 0$ for the triplet state, its x-, y- and z-components must each be zero, and the individual nucleon spins are genuinely antiparallel, component by component, so that $\boldsymbol{\sigma}_1 \cdot \boldsymbol{\sigma}_2 = -\sigma_1^2 = -3$.

The loose statement that in the triplet state the nucleon spins are parallel has no such component-by-component interpretation, since the three components of a non-zero spin are not simultaneously observable. If the z-component is known, the only thing we know about the x and y components is the eigenvalue of $S_x^2 + S_y^2$:

$$\langle S_x^2 + S_y^2 \rangle = \langle \mathbf{S}^2 \rangle - \langle S_z^2 \rangle$$

$$= \begin{cases} 2 & \text{for the centre member of the triplet (with } S_z = 0\text{),} \\ 1 & \text{for the extreme members of the triplet (with } S_z = \pm 1\text{).} \end{cases}$$

Here, $\mathbf{r}$ is the vector describing the position of the proton with respect to the neutron, and $\boldsymbol{\sigma}_1$ and $\boldsymbol{\sigma}_2$ are again the Pauli operators representing the spins of the two nucleons. The term $\boldsymbol{\sigma}_1 . \boldsymbol{\sigma}_2$ in S_{12} is included to make the average over all directions of $\mathbf{r}$ vanish.

Combining equations **12.3** and **12.4**, one obtains, as the most general form for the total nucleon–nucleon potential, depending on positions and spins but not on velocities,

$$V = V_d(r) + V_\sigma(r)\,\boldsymbol{\sigma}_1 . \boldsymbol{\sigma}_2 + V_T(r)S_{12}.$$

All three contributions to the potential are dependent on the radius r, though not necessarily in the same way.

In the singlet state, where there is no preferred direction, we have

$$S_{12} = 0$$

and the forces can only be central.

But in the triplet spin-state, the tensor force can lead to an admixture of a state with $l = 2$ in the neutron–proton system. This small D-state term in the wave function can interfere with the predominant S-state term to give a non-spherical distribution of charge. If the charge distribution is described in terms of an axially symmetrical probability distribution for a point proton of charge e, the quadrupole moment is

$$Q = e . \tfrac{1}{4}\overline{(3z^2 - r^2)},$$

where the bar indicates mean value. The method of relating this quantity to the magnitude of the D-state term, and to the strength of the tensor force, will be found in the literature (e.g. Blatt and Weisskopf, 1952, pp. 95–108).

12.4 Polarization experiments

12.4.1 *Polarization by scattering*

As has been mentioned in section 12.2, the shell model of the nucleus requires a strong spin–orbit coupling for nucleons inside nuclei. By extension of this idea to unbound states, in which a nucleon is undergoing scattering as it passes outside a target nucleus, we may expect an interaction between the spin of the nucleon and the orbital angular momentum of its motion relative to the target, provided that it passes close enough to be within range of the spin-dependent nuclear force.

Let us consider elastic scattering of a nucleon by a heavy target nucleus (assumed to have zero spin for simplicity). On a classical model the orbital angular momentum is

$$l = \mathbf{r} \times \mathbf{p},$$

where $\mathbf{p}$ is the momentum of the incident nucleon and $\mathbf{r}$ its position relative to the target. If we imagine the motion to be in a horizontal plane, the vector

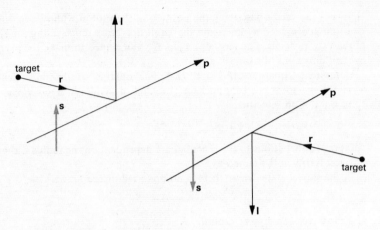

Figure 52 Orbital angular momentum and spin–orbit interaction

l will be up when the nucleon is passing to the right of the target, and down when it is passing to the left (see Figure 52). If the spin of the nucleon is a vector s, the energy of the spin–orbit interaction will be proportional to the scalar product $s \cdot l$ (this is a true scalar, because both s and l are pseudovectors, or axial vectors, whose sense depends on a conventional choice of right-handed axes). So we expect a nucleon with spin up passing to the right of the target to have the same spin–orbit interaction as a nucleon with spin down passing to the left. Opposite spin–orbit interaction will occur when a nucleon with spin up passes to the left, or a nucleon with spin down passes to the right.

The differential cross-section for scattering will depend on the total energy of interaction of the nucleon with the target, and the scattering will be to right or left according to which side of the target the nucleon passes. Thus there will be one differential cross-section for right-handed scattering of nucleons with spin up, or left-handed scattering of nucleons with spin down ($\sigma_{R+} = \sigma_{L-}$), and another differential cross-section for left-handed scattering with spin up, or right-handed scattering with spin down ($\sigma_{L+} = \sigma_{R-}$). It follows that when an unpolarized beam of protons or neutrons is scattered in a target, the particles scattered to one side will be predominantly those with spin up, and those scattered to the other side predominantly with spin down. (Which side is which will depend on the sign of the effect for the particular scattering process.) The extent of the predominance at a particular angle will depend upon the difference between the two differential cross-sections for that angle.

The polarization of a beam of particles of spin one half is defined as

$$P = \frac{N_+ - N_-}{N_+ + N_-},$$

where N_+ is the number of particles with z-component of spin $+\frac{1}{2}$ (i.e. spin up), and N_- is the number with z-component of spin $-\frac{1}{2}$ (i.e. spin down). P can have values ranging from $+1$ (completely polarized upwards), through 0 (unpolarized) to -1 (completely polarized downwards). For the present work, the z-axis is taken perpendicular to the plane of scattering of the particle. Thus we are considering transverse polarization only.

After a scattering to the right, N_+ will be proportional to σ_{R+}, and N_- to σ_{R-}, so the particles scattered through an angle θ to the right will have a polarization

$$P(\theta) = \frac{N_+ - N_-}{N_+ + N_-} = \frac{\sigma_{R+}(\theta) - \sigma_{R-}(\theta)}{\sigma_{R+}(\theta) + \sigma_{R-}(\theta)}. \qquad \textbf{12.8}$$

This is a function of θ, which we have seen may be non-zero if some of the scattering occurs in partial waves with $l \neq 0$.

Thus a necessary condition for obtaining polarization by scattering from a spinless target is that the energy should be high enough to allow some P-wave scattering. The table in section 5.4.3 shows that P-wave scattering does not become important for neutrons until the incident kinetic energy reaches the order of 100 MeV for collisions with protons, or 5 MeV for collisions with carbon nuclei. With incident protons, some P-wave scattering can occur at lower energies, through the long-range Coulomb interaction. But this interaction is predominantly spin independent, so polarization of protons is subject to roughly the same limitations of energy as is that of neutrons.

A particularly favourable case, in which the amplitude of P-wave scattering is increased by resonance scattering, was pointed out by Wolfenstein (1949). This is the case of neutron scattering by helium-4, through the broad resonance of $J = \frac{3}{2}$ which corresponds to formation of the unstable nucleus helium-5 in its ground state. The strong spin dependence of the interaction is reflected in the fact that the level in helium-5 with $J = \frac{1}{2}$ is 2–3 MeV above the ground state. Thus, while P-wave n–α scattering in the state with neutron spin and orbital angular momentum parallel has a resonance at about 1 MeV, scattering in the state where they are antiparallel does not resonate until about 4 MeV.

A very similar situation occurs in the scattering of protons by helium, which has a resonance at about 2 MeV, where the ground state of lithium-5 is formed. Experimental study of the polarization showed (Heusinkveld and Freier, 1952) that the ground state was indeed the one with $J = \frac{3}{2}$, and that P changed sign at 4 MeV as it should if the level of $J = \frac{1}{2}$ was at a higher energy (Juveland and Jentschke, 1956). Measurements by Scott (1958) showed that the polarization reached a peak value of 85 per cent at 2·0 MeV.

The proper quantum-mechanical treatment (see e.g. Fernbach, Heckrotte and Lepore, 1955) shows quantitatively that the observed polarizations, of order 50 per cent and occasionally higher, can indeed result from spin–orbit interaction in scattering of particles passing close to the nuclear surface. While the classical model requires only an interaction energy dependent on $\mathbf{s} \cdot \boldsymbol{l}$, the quantum-mechanical model describes an asymmetry in terms of interference between partial waves of opposite parity, for example, between the P-wave of which the orbital angular momentum is coupled to the spin, and the background of S-wave scattering which on its own would be spin independent.

12.4.2 *Measurement of polarization*

If a scattering process is capable of producing polarization, it can also be used for the detection and measurement of polarization.

Figure 53 Double scattering

When a beam of nucleons has acquired a polarization P by a first scattering through an angle θ to the right, a second scattering through the same angle in a second similar target gives a left–right asymmetry. The asymmetry ε is defined in terms of the numbers R and L scattered to right and left in the second target of the double-scattering experiment illustrated in Figure 53 by

$$\varepsilon = \frac{R-L}{R+L}. \qquad\qquad \textbf{12.9}$$

If the two scatterings occur at effectively the same energy as well as the same angle, the same cross-sections may be used for them both, as follows.

The relative numbers of particles hitting the second target with spin up and spin down are given by the numbers N_+ and N_- leaving the first target. They are in the ratio σ_{R+}/σ_{R-}, which is related to the polarization P by equation **12.8**. The total number of particles scattered to the right in the second target is

$$R = c_1(N_+ \, \sigma_{R+} + N_- \, \sigma_{R-})$$
$$= c_2(\sigma_{R+}^2 + \sigma_{R-}^2).$$

The number scattered to the left in the second target is

$$L = c_1(N_+ \, \sigma_{L+} + N_- \, \sigma_{L-})$$
$$= c_1(N_+ \sigma_{R-} + N_- \, \sigma_{R+}) = c_2 . 2\sigma_{R+} \, \sigma_{R-}.$$

L differs from R because L represents the number scattered first to the right and then to the left, while R is the number undergoing two successive scatters to the same side. The constants c_1 and c_2 are geometrical factors involving target separation, sizes and densities.

The important feature in these results is the ratio R/L, and the asymmetry ε, which equation **12.9** now gives as

$$\varepsilon = \frac{R-L}{R+L} = \frac{\sigma_{R+}^2 + \sigma_{R-}^2 - 2\sigma_{R+} \, \sigma_{R-}}{\sigma_{R+}^2 + \sigma_{R-}^2 + 2\sigma_{R+} \, \sigma_{R-}}$$

$$\varepsilon = \left[\frac{\sigma_{R+} - \sigma_{R-}}{\sigma_{R+} + \sigma_{R-}} \right]^2 = P^2. \qquad \textbf{12.10}$$

Thus the asymmetry following a second scatter is equal to the square of the polarization produced in a single scattering, and may be used to determine the magnitude of P, though not its sign.

If the scatterings are not similar, through being at different energies or angles, or in different types of target, the asymmetry is

$$\varepsilon = P_1 P_2, \qquad \textbf{12.11}$$

where P_1 and P_2 are the polarizations which the two types of scatter would have caused separately, in single-scattering experiments.

Thus a beam polarized by a first scattering may be used to study the spin dependence of interactions occurring in the second target.

Alternatively, and more commonly, the second scattering may be used as an analysing process for the study of the interactions involved in the first scattering; P_2 may be called its analysing power. This is in accordance with the general relation

$$\varepsilon = \alpha P, \qquad \textbf{12.12}$$

in which α is the analysing power of any analysing process, which may be scattering, preferential absorption in a magnetized target, or parity-violating decay in the case of an unstable particle.

In practice, scattering in helium or carbon is widely used for analysing the polarization of proton and neutron beams.

12.4.3   *Triple scattering experiments*

A complete phenomenological description of the spin dependence of an interaction includes not only the differential cross-section as a function of the initial spin direction, as discussed above, but also the distribution of final spin directions. The latter is described by the three Wolfenstein parameters R, D and A (see Wolfenstein, 1956, p. 51, or Segré, 1964, p. 413).

D gives the depolarization resulting from a scattering in the plane perpendicular to the existing polarization. R, for rotation, gives the transverse polarization after a scattering in the plane containing the initial polarization. A gives the transverse polarization, in the plane of scattering, resulting from an initial longitudinal polarization.

To measure these quantities we need triple scattering experiments in which the first scatter produces a polarized beam, the second is the process under examination, and the third serves to analyse the final polarization.

For D, we need three successive scatterings in the same plane (see Figure 54a). For a spinless target, there is no depolarization. For R, the polarizing and analysing scatters are in the same plane but this is perpendicular to that of the second scatter (see Figure 54b).

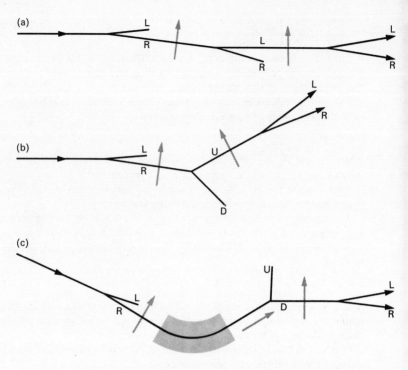

Figure 54 Triple scattering experiments. (a) For *D*, which relates transverse polarizations before and after a scatter in the plane perpendicular to the polarization. (b) For *R*, which gives transverse polarization after a scatter in the plane of an initial transverse polarization. (c) For *A*, which gives transverse polarization after scattering of longitudinally polarized particles

A (see Figure 54c) needs the same layout as *R*, with a magnet between the first and second scatters, to convert transverse to longitudinal polarization (see section 12.4).

With experimental values of $d\sigma/d\Omega$ from a single-scattering experiment, *P* from a double-scattering experiment, and *R* and *A* from triple scattering, we would have all the information needed for calculating a complete description of scattering of nucleons at a single angle by a spinless target. The description is usually given in terms of a matrix scattering amplitude

$$M = g(\theta) + h(\theta)\boldsymbol{\sigma} \cdot \mathbf{n}.$$

Here **n** is the unit vector perpendicular to the plane of scattering, $\boldsymbol{\sigma}$ has for its three components the three Pauli spin matrices for the nucleon, $g(\theta)$ is the ordinary scattering amplitude for scattering without change of spin direction, and $h(\theta)$ is known as the spin-flip scattering amplitude. This name, though expressive, is not wholly accurate, for $h(\theta)$ causes the nucleon spin to change direction only when it is in the plane of scattering. When the spin is perpendicular to the scattering plane, $h(\theta)$ contributes to the scattering without causing the spin to flip over.

12.4.4 *Nucleon–nucleon scattering*

Proton–proton and neutron–proton scattering both show marked polarization effects at energies high enough to give P-wave scattering; most of the experiments are in the interval from 100 MeV to 500 MeV (see review by Wolfenstein, 1956). The simple model of spin–orbit coupling, used above for a spinless target, is not adequate for the case of a particle of spin one half scattering off a target particle of spin one half. The central force due to interaction between the two spins produces no polarization, but the non-central tensor force gives important effects, which have to be included in any complete analysis.

In this case, the description of what happens in terms of $d\sigma/d\Omega$, P, D, R and A is incomplete, for we need also the final polarization of the second particle, and the correlations between the two spin directions. Thus the case when the target and scattered particle both have spin is beyond the scope of this book, not only in interpretation, but also in the mere description of experimental fact. The full specification of even proton–proton scattering at a single energy and angle would require measurement of nine independent quantities (Wolfenstein, 1956, pp. 67–8).

12.5 **Polarized beams**

12.5.1 *Sources of polarized particles*

Beams of polarized particles may be set up in many ways, of which some are summarized in the following list.

Scattering. As was discussed in section 12.4, elastically scattered particles are in general polarized transversely, in a direction perpendicular to the plane of scattering. This is a simple way of obtaining polarized protons and neutrons, but it suffers limitation in the matter of intensity.

Nuclear reactions. The products of nuclear reactions may be polarized through spin–orbit coupling in the interactions by which they are formed (elastic scattering is really a special case of this). Examples of especial interest are the reactions $^7\text{Li(p, n)}^7\text{Be}$ and $^3\text{He(d, n)}^4\text{He}$, both of which may be used to produce beams of high-energy neutrons with about 50 per cent polarization.

Reflection by magnetized iron. Slow neutrons from a nuclear reactor can be almost completely polarized by Bragg reflection from a magnetized ferromagnetic material. The polarization results from the spin dependence of the scattering amplitude for scattering of the neutrons by the unpaired electrons, which are aligned in a magnetized ferromagnetic material.

215   Polarized beams

Polarizing ion sources. Protons or deuterons to be accelerated in a linear or circular accelerating machine are usually produced by an electric discharge in an appropriate ion source. Sources can be designed to produce polarized protons or deuterons, which in suitable conditions retain their polarization through the accelerating process, yielding a beam of high-energy polarized particles. 'Suitable conditions' means, in practice, absence of strongly focusing magnetic fields that would cause depolarization by precession of individual particle spins through angles dependent on the path followed.

This technique offers a satisfactory way of obtaining intense beams of polarized charged particles, under well-controlled conditions. With such a beam, double scattering is enough to give information of the type provided by ordinary triple scattering experiments. By cutting out the loss of intensity in the first of three scatterings, it is possible to extend enormously the range of measurements which can be made, studying otherwise inaccessible processes, and accessible ones with greatly improved statistics.

To provide a basis for discussing sources of polarized protons, we now discuss the states of the hydrogen atom.

12.5.2   *States of the hydrogen atom*

The hydrogen atom, in a region of zero magnetic field, has two states, the lower with total spin zero (i.e. with electron and proton spins antiparallel), and the higher with total spin one. These two states are represented by the left-hand ends of the plots shown in Figure 55, where they are labelled S (singlet) and T (triplet).

In a moderately strong magnetic field H there are four states, with energies depending on H as shown by the right-hand parts of the four lines in Figure 55. The largest contribution to the energy differences between these states comes from the interaction of the magnetic moment of the electron with the applied magnetic field H. This gives a positive energy when the magnetic moment is antiparallel to H, that is, when the electron spin is parallel to H. The electron in turn makes a magnetic field of order 10^5 G at the position of the proton, in the same sense as the magnetic moment of the electron. Thus the second contribution to the total energy of the atom comes from the orientation of the proton magnetic moment with respect to that of the electron, and is positive when the two magnetic moments are antiparallel. For all but the highest attainable values of H, the direct interaction of the proton magnetic moment with the external magnetic field is swamped by their interaction via the electron. The orientations of the two magnetic moments in the four states are shown in Figure 55, as are those of the two spins (parallel to the magnetic moment for the proton, but opposite for the electron).

From the spin directions shown in Figure 55 it will be seen that the highest state (state 1) and the third have total z-component of spin ± 1, and hence

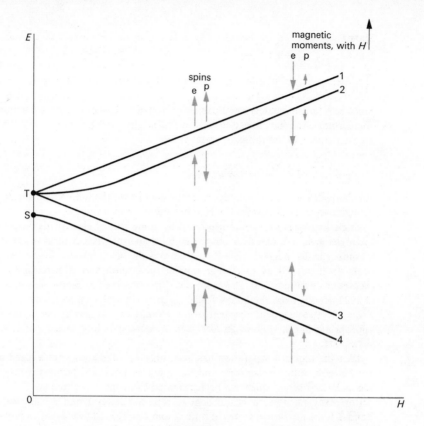

Figure 55 States of the hydrogen atom

spin 1. States 2 and 4 both have total z-component of spin zero. They are eigenstates of the system at high magnetic field, but as the field is reduced the eigenstates become mixtures of states 2 and 4, until at zero field they become the symmetric and antisymmetric fifty-fifty mixtures T and S. The energies for these eigenstates depend nonlinearly on H, and are plotted in Figure 55 as the curves connecting the straight line for state 2 with the point representing T, and the straight line for state 4 with the point for S. Thus T really represents the three triplet states, which are degenerate at zero magnetic field, and the line leaving T horizontally represents the middle one of these three triplet states, with zero z-component of total spin, and energy independent of field at small fields. But as the field increases, this line represents a changing mixture of states, becoming straight when it represents the pure state 2.

The straight parts of the lines in Figure 55 have slopes given by the magnetic moments μ_e and μ_p of the electron and the proton as follows:

State 1 $|\mu_e| - \mu_p$,

State 2 $|\mu_e| + \mu_p$,

State 3 $-|\mu_e| + \mu_p$,

State 4 $-|\mu_e| - \mu_p$.

Since μ_e is approximately one Bohr magneton, and μ_p is only 2·79 of the much smaller nuclear magnetons (see section 12.1.4), the plots for states 1 and 2 are nearly parallel, as are those for states 3 and 4.

12.5.3   *The principle of the polarizing source*

Virtually all sources of polarized protons depend on the following principles, though they use different methods of putting them into practice.

Single hydrogen atoms are produced by some sort of electric discharge in hydrogen gas, and then flow out through an aperture as a neutral beam. An inhomogeneous magnetic field then separates the beam into the four components distinguished by the orientation of the proton and electron spins in each atom (see Figure 55). This separation, first observed by Stern and Gerlach in 1922, results from the lateral force μ_{eff} grad $|\mathbf{H}|$ on each atom, where μ_{eff} is the effective magnetic moment of the atom, and $-\mu_{eff}$ is the slope of the appropriate line in Figure 55 at the point corresponding to the particular value of $\mathbf{H}$.

After the required separation has been obtained, the atoms are ionized by a transverse beam of electrons, and the resultant polarized protons continue as a polarized beam which can be accelerated by ordinary techniques.

For many purposes it is sufficient to separate states 1 and 2 from states 3 and 4. If the mixture of atoms in states 1 and 2 is allowed to pass into a region of small magnetic field, those in state 1 will retain their orientation with respect to the field; but those in state 2, having small μ_{eff} at small field, will lose their orientation. The resultant mixture of atoms in state 1 with protons aligned, and equal numbers of unaligned atoms in state 2, enters the ionizer and yields a proton beam with 50 per cent polarization.

In practice the inhomogeneous field is usually provided by a quadrupole or a sextupole magnet. These give fields which increase with distance from the axis, so that atoms with positive μ_{eff} (states 3 and 4) experience defocusing forces away from the axis, while those with negative μ_{eff} (states 1 and 2) are focused by forces directed towards the axis. Suitable apertures ensure that the defocused components are lost.

The field itself is in a tangential direction, so that the focused atoms of states 1 and 2 emerge from the quadrupole or sextupole magnet with their orientations distributed over the transverse plane. They then move smoothly into the weaker, homogeneous field produced for example by a pair of Helmholtz coils. During this transition, the atoms of state 1 take up a new orientation along the homogeneous field, but those of state 2 end with random orientations.

To get more than 50 per cent polarization, or to allow the ionizer to work in a high magnetic field (e.g. at the centre of a cyclotron), states 1 and 2 must be separated from each other. This may be achieved by a further carefully designed inhomogeneous magnetic field, with especially good effect if the magnitude of the field is allowed to drop to about two hundred gauss towards the end of the path. In such a field the effective magnetic moment of state 2 is already much less than that of state 1, so the transverse forces on atoms in the two states are significantly different.

An alternative method is to depopulate state 2 by using a radio-frequency field to induce transitions from state 2 to state 4. This refinement brings the polarizing ion source nearer to the atomic beam experiments discussed in section 12.2.6.

12.5.4   *The construction of polarizing ion sources*

The principles and practice of polarizing ion sources are discussed at length in the literature (see, e.g., the reviews by Dickson (1965) and Daniels (1965, chapter 6). Among actual sources we may pick out those constructed at CERN (Keller, Dick and Fidecaro, 1961), at Minnesota (Clausnitzer, 1963) and at the Rutherford High Energy Laboratory (Stafford et al., 1962). Of these the last two use sextupole magnets to select states 1 and 2 without separating them from each other. None of the three uses any radio-frequency transition to reduce the population of state 2, but the CERN source uses a seven-metre tapered quadrupole magnet to separate state 1 from state 2. This is necessary since the atomic beam is carried to the centre of the CERN synchrocyclotron where it is ionized in a magnetic field of 18 000 G, prior to acceleration up to 600 MeV. In the Minnesota and Rutherford Laboratory sources, ionization occurs in fields of about 20 G, small enough to ensure that atoms in state 2 are randomly oriented. Linear accelerators then accelerate the protons to kinetic energies of 68 and 50 MeV respectively. Polarizations up to three-quarters of the theoretical maximum (50 per cent) have been observed in the final beams.

Transverse polarization is usually wanted, and its sense may be selected by choosing the direction of the current in the Helmholtz coils. After acceleration, magnetic deflection in the plane perpendicular to the polarization leaves it unaffected. If longitudinal polarization should be needed, it may be obtained by the following device. A solenoid between the ionizer and the accelerator causes precession of the transverse polarization through 90°. According to whether this solenoid is on or off, the accelerated beam enters a bending magnet polarized in the plane of bending or at right angles to it. In the latter case, the polarization remains transverse; but in the former, a bending through 50° causes precession of the polarization through $2 \cdot 79 \times 50° = 140°$, which is sufficient to change transverse into longitudinal polarization†. Thus a fixed path, including a bending magnet giving a 50° deflection, can be used to

---

†See section 12.1.8 on $g - 2$ experiments, in which the change from transverse to longitudinal polarization needs many complete circular orbits.

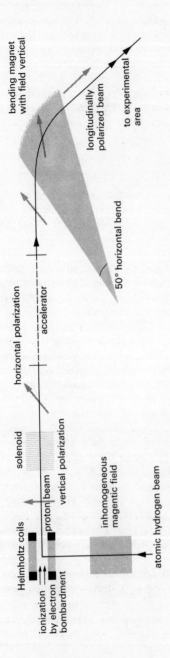

Figure 56 Schematic layout of a polarized proton beam. The direction of current in the Helmholtz coils determines whether the initial polarization of the proton beam is up or down. If the solenoid is off, the initial transverse polarization is maintained, unaffected by the bending magnet, and the protons reach the experimental area vertically polarized. With the solenoid on, the protons reach the bending magnet with horizontal polarization. The 50° bend in the horizontal plane causes 140° precession, leading to longitudinal polarization as shown

deliver either transversely or longitudinally polarized protons. A layout for achieving this is illustrated in Figure 56.

12.6 Nuclear polarization in solids

12.6.1 *Conditions for polarization*

Normally the nuclei of atoms in a solid are randomly oriented. If they are to be polarized, the following conditions must be satisfied:

(a) A direction for the possible polarization must be fixed by an externally applied magnetic field.

(b) A mechanism for coupling the nuclear magnetic moments to the magnetic field, either directly or indirectly, must be provided.

(c) By one means or another, the states distinguished by this mechanism must be differently populated.

(d) The temperature must be low enough to prevent any polarization from being destroyed by thermal redistribution of the populations of the different states.

In one class of method, condition (d) is used to provide the differential population required in (c). Thermal equilibrium is allowed to depopulate the higher of two levels which are separated by an energy greater than $k\Theta$. Methods of this type are discussed in section 12.6.2.

In another broad class of method, a non-thermal distribution of populations is obtained by using microwaves to induce transitions to levels in which the nuclei are polarized, or achieve a net polarization in some relaxation process. An example of this type of method is described in section 12.6.3.

12.6.2 *Polarization through thermal equilibrium*

The direct method. If a high magnetic field H is applied directly to a solid, different orientations of the nuclei will lead to a hyperfine splitting of the energy levels. The energy of this splitting is of the order $\mu_N H$, and if it is to result in any net polarization of the nuclei through preferential population of the lower levels, the temperature must be low enough to make

$$\mu_N H \gtrsim k\Theta.$$

The nuclear magneton μ_N is $3 \cdot 1524 \times 10^{-12}$ eV G^{-1}, and the Boltzmann constant k is $8 \cdot 6171 \times 10^{-5}$ eV K^{-1}, so this condition appears to require

$$\Theta \lesssim 3 \cdot 7 \times 10^{-3} \text{ K} \quad \text{if } H \simeq 10^5 \text{ G}.$$

In many cases this very low value for Θ is a true reflection of the difficulty of obtaining nuclear polarization through thermal equilibrium following direct interaction between the nuclei and an external field. But when the nucleus has a magnetic moment of several nuclear magnetons (say $\mu \simeq 4$), the total separation between the highest and lowest hyperfine level is $2\mu\mu_N H$, which improves the picture by almost an order of magnitude. Even so, one needs a magnetic field approaching 10^5 G and a temperature about 0·01 K. Small polarizations have been obtained in less rigorous conditions, for example, in copper at 28 kG at a temperature of 0·01 K (Kurti *et al.*, 1956).

Methods using paramagnetic ions. The magnetic field at the nucleus of a paramagnetic ion, due to the unpaired atomic electron, is of order 10^5 G, and an externally applied field very much smaller than this is sufficient to polarize such ions at a temperature about 1 K. It follows that the electrons of paramagnetic ions can provide a useful indirect means of coupling nuclear magnetic moments to an applied magnetic field, leading to a greater hyperfine splitting than could be obtained directly. Thermal equilibrium at about 1 K can then lead to appreciable nuclear polarization. Alternatively, adiabatic demagnetization may be used to reduce the temperature to about 0·01 K, and a higher equilibrium polarization obtained.

An important application of this method was the experiment of Wu *et al.* (1957) demonstrating the parity-violating beta decay of the cobalt-60 nucleus. For this experiment, the salt $Ce_2Mg_3(NO_3)_{12}.24H_2O$ was used, with some of the magnesium replaced by cobalt-60. Earlier work had shown that polarizations of 50 per cent could be obtained in this material with an external field of about five hundred gauss, after cooling to 0·004 K by adiabatic demagnetization.

12.6.3 *Polarization by microwave pumping*

In the second class of method, equilibrium which is not thermal is obtained by using microwaves to 'pump' atoms from one level to another. This method was pioneered by Jeffries (1960), and by Abragam and Proctor (1958) and Abragam *et al.* (1962). It has since been the basis for polarized proton targets for many experiments in nuclear and elementary particle physics (see Leifson and Jeffries, 1961, Duke *et al.*, 1965).

A suitable material is $La_2Mg(NO_3)_{12}.24H_2O$, with neodymium replacing 1 per cent of the lanthanum. The paramagnetic Nd^{3+} ions are loosely coupled to the protons in the water of crystallization, but the strongest couplings are of each of these to an external magnetic field H, so that the sequence of levels is as shown in Figure 57.

At a temperature of about 1·2 K, the levels A and B are roughly equally populated, but the coupling of the paramagnetic Nd^{3+} ions to the field raises levels C and D well above $k\Theta$. Radio-frequency excitation at a frequency corresponding to the energy of the transition $B \leftrightarrow C$ can now equalize the populations of states B and C. From state C, relaxation to state A occurs rapidly, so a steady state is reached with state A the most fully populated and states B and C relatively empty. The excess of A and C over B in practice gives a proton polarization of about 60 per cent.

An attractive feature of these polarized proton targets is the possibility of reversing their polarization without changing the magnetic field. This can be

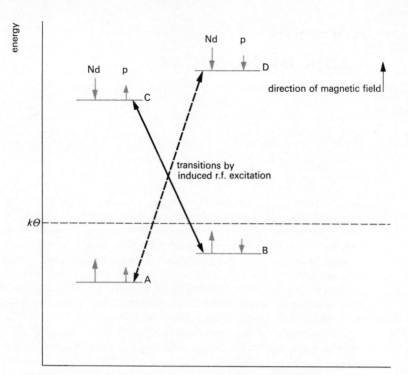

Figure 57 Levels in $La_2Mg(NO_3)_{12}$. $24H_2O$ (1 per cent lanthanum replaced by neodymium), with $k\Theta$ at temperature ($\Theta \sim 1 \cdot 2K$) used for polarized proton target

left untouched, with trajectories of particles in and out of the target unaltered, while the microwave frequency is changed to give transitions $A \leftrightarrow D$ instead of $B \leftrightarrow C$.

Thus instead of measuring a left–right asymmetry with supposedly equivalent counters on the two sides, one can obtain equivalent information, free from the main systematic errors, from a counter system on one side only, the polarization of the target being changed to make this side either left or right with respect to the polarized protons.

The high density of the materials used for these targets makes them more suitable for the study of high-energy processes than low-energy. In the former field they have provided critical information, especially for the allocation of parity and spin quantum numbers to nucleon resonances excited in pion–nucleon scattering.

To conclude this discussion, one should point out that the spin dependence of nuclear interactions still offers a wide-open field for exploration with the developing techniques available for polarization of beams and targets.

Appendix
Table of Nuclides

Table of Nuclides

The table presented in this appendix summarizes the main properties of known types of nuclei in their ground states. In deciding which isotopes of a given element should be included, the rule has been to include all those whose existence is well established, stopping at the point where the experimental information begins to indicate a significant element of doubt. Thus many well-established but incompletely documented isotopes are included, with blanks for their unmeasured properties. However, a subjective criterion of this type must lead to some errors and inconsistencies, and no canonical significance should be given to the individual applications of it.

The mass number A and mass excess $M - A$ are listed in adjacent columns, in the same units, so that the true atomic masses M may be read directly by adding the pairs of entries. The mass unit used is $\frac{1}{12}$ of the mass of a carbon-12 atom, which is equivalent to an energy of 931·44 MeV. As is customary in such tables, the masses are atomic masses, not nuclear masses. The atomic mass is taken to be the nuclear mass plus that of Z free electrons.

The binding energy is the total stored energy relative to that of Z protons ($+Z$ electrons) plus $A - Z$ neutrons. Values for both mass excess and binding energy are taken from the tables of Mattauch, Thiele and Wapstra (1965).

The values listed for nuclear spin, when known, and for the nuclear magnetic moment in nuclear magnetons, are experimental values taken from Strominger, Hollander and Seaborg (1958) and from the *Nuclear Data Sheets* of Way *et al.* (n.d.), as are the quoted values for abundance of stable or long-lived isotopes, and for the decay modes and energies of unstable isotopes.

For the more recently discovered transuranic elements, however, reference has been made to the more specialized reviews such as Seaborg (1968), and in some cases to the original publications (for fermium, to Nurmia *et al.*, 1967; for mendelevium, to Silkeland *et al.*, 1965; for Rf and Ha, to Ghiorso *et al.*, 1969, 1970).

Element	Z	A	$(M-A)$/a.m.u.	Binding energy/ MeV	Spin	Magnetic moment	Percentage abundance	halflife	Decay mode	energy/MeV
n	0	1	+0·008 665	0	$\frac{1}{2}$	−1·913		10·8 m	e^+	0·783
H	1	1	+0·007 825	0	$\frac{1}{2}$	2·973	99·986			
		2	+0·014 102	2·225	1	0·857	0·014			
		3	+0·016 049	8·482	$\frac{1}{2}$	2·979		12·3 y	e^-	0·0181
He	2	3	+0·016 029	7·718	$\frac{1}{2}$	−2·127	10^{-4}			
		4	+0·002 603	28·296	0		~100			
		5	+0·012 297	27·338						
		6	+0·018 893	29·266	0			0·82 s	e^-	3·515
Li	3	5	+0·012 538	26·331						
		6	+0·015 125	31·993	1	0·882	7·4			
		7	+0·016 004	39·245	$\frac{3}{2}$	3·256	92·6			
		8	+0·022 487	41·278	2			0·85 s	e^-	16·0
Be	4	7	+0·016 929	37·601	$\frac{3}{2}$			53·6 d	EC	0·86
		8	+0·005 308	56·497				~10^{-15} s	2α	
		9	+0·012 186	58·163	$\frac{3}{2}$	−1·177	100			
		10	+0·013 534	64·978	0			$2·5 \times 10^6$ y	e^-	0·555
B	5	8	+0·024 609	37·736				0·8 s	e^-	13·7
		9	+0·013 332	56·312						
		10	+0·012 939	64·750	3	1·801	19			
		11	+0·009 305	76·205	$\frac{3}{2}$	2·688	81			
		12	+0·014 353	79·574	1			0·02 s	e^-	13·4

Element	Z	A	(M−A)/a.m.u.	Binding energy/MeV	Spin	Magnetic moment	Percentage abundance	halflife	Decay mode	Decay energy/MeV
C	6	10	+0·016 810	60·361	0			19·1 s	e+	2·1
		11	+0·011 432	73·443	$\frac{3}{2}$			20 m	e+	0·99
		12	0	92·163	0		98·9			
		13	+0·003 354	97·109	$\frac{1}{2}$	0·702	1·1			
		14	+0·003 242	105·286	0			5600 y	e−	0·155
N	7	12	+0·018 641	74·017	1			0·0125 s	e+	16·6
		13	+0·005 738	94·106	$\frac{1}{2}$	±0·322		10·0 m	e+	1·2
		14	+0·003 074	104·659	1	0·403	99·635			
		15	+0·000 108	115·494	$\frac{1}{2}$	−0·283	0·365			
		16	+0·006 103	117·981	2			7·3 s	e−	10·4
		17	+0·008 450	123·867				4·1 s	e−	8·8
O	8	14	+0·008 597	98·732	0			72 s	e+	4·1 & 1·8
		15	+0·003 070	111·952	$\frac{1}{2}$	±0·719		124 s	e+	1·72
		16	−0·005 085	127·620	0		99·759			
		17	−0·000 867	131·763	$\frac{5}{2}$	−1·893	0·037			
		18	−0·000 840	139·809	0		0·204			
		19	+0·003 578	143·765	$\frac{5}{2}$			29 s	e−	4·5 & 2·9
F	9	17	+0·002 096	128·220	1			66 s	e+	1·75
		18	+0·000 937	137·371	$\frac{1}{2}$			112 m	e+	0·65
		19	−0·001 595	147·801	2	2·628	100			
		20	−0·000 013	154·399	2			11 s	e−	5·4

Element	Z	A			spin		%	half-life		
Ne	10	18	+0.005 711	132·142	0			1·6 s	e^+	3·2
		19	+0.001 881	143·781	$\frac{1}{2}$	−1·887		18 s	e^+	2·23
		20	−0.007 560	160·646	0		90·92			
		21	−0.006 151	167·406	$\frac{3}{2}$	−0·661	0·26			
		22	−0.008 615	177·772	0		8·82			
		23	−0.005 527	182·967				40 s	e^-	4·4
		24	−0.006 387	191·839				3·4 m	e^-	2·0
Na	11	20	+0.008 880	144·550				0·385 s	e^+	$Q = 14·3$
		21	−0.002 345	163·078	$\frac{3}{2}$	2·386		23 s	e^+	2·50
		22	−0.005 563	174·147	3	1·746		2·6 y	e^+	1·8 & 0·54
		23	−0.010 229	186·564	$\frac{3}{2}$	2·217	100			
		24	−0.009 038	193·526	4	1·69		15·0 h	e^-	1·4 & 4·2
		25	−0.010 045	202·535	$\frac{5}{2}$			60 s	e^-	4·0
Mg	12	23	−0.005 875	181·726	$\frac{3}{2}$			12 s	e^+	2·95
		24	−0.014 958	198·258	0		78·7			
		25	−0.014 161	205·587	$\frac{5}{2}$	−0·855	10·1			
		26	−0.017 407	216·682	0		11·2			
		27	−0.015 655	223·122	$\frac{1}{2}$			9·4 m	e^-	1·8
		28	−0.016 125	231·631	0			22 h	e^-	0·4
Al	13	24	+0.000 100	183·450				2·1 s	e^+, γ	$Q = 13·0$
		25	−0.009 588	200·545	$\frac{5}{2}$			7·6 s	e^+	3·24
		26	−0.013 109	211·896	5			8×10^5 y	e^+	1·2
		27	−0.018 461	224·953	$\frac{5}{2}$	3·641	100			
		28	−0.018 095	232·684	3			2·3 m	e^-	2·8
		29	−0.019 558	242·118				6·6 m	e^-	2·5

Element	Z	A	$(M-A)$/a.m.u.	Binding energy/ MeV	Spin	Magnetic moment	Percentage abundance	halflife	Decay mode	Decay energy/MeV
Si	14	27	−0·013 297	219·360	$\frac{5}{2}$			4·4 s	e^+	3·7
		28	−0·023 071	236·536	0		92·2			
		29	−0·023 504	245·011	$\frac{1}{2}$	−0·555	4·7			
		30	−0·026 237	255·628	0		3·1			
		31	−0·024 651	262·222	$\frac{3}{2}$			2·6 h	e^-	1·5
		32	−0·025 980	271·530	0			710 y	e^-	0·1
P	15	28	−0·008 220	221·920	$\frac{1}{2}$			0·28 s	e^+	10·6
		29	−0·018 192	239·280	1			4·5 s	e^+	3·95
		30	−0·021 683	250·603	1			2·5 m	e^+	3·24
		31	−0·026 235	262·915	$\frac{1}{2}$	1·130	100			
		32	−0·026 090	270·852	1	−0·252		1·4 d	e^-	1·70
		33	−0·028 272	280·955				25 d	e^-	0·25
		34	−0·026 660	287·530	1			12·5 s	e^-	5·1
S	16	31	−0·020 388	256·689	$\frac{1}{2}$			3·0 s	e^+	4·5
		32	−0·027 926	271·780	0		95			
		33	−0·028 538	280·421	$\frac{3}{2}$	0·643	0·75			
		34	−0·032 135	291·843	0		4·23			
		35	−0·030 969	298·828	$\frac{3}{2}$	±1·035		87 d	e^-	0·17
		36	−0·032 910	308·707	0		0·02			
		37	−0·028 990	313·130				5·0 m	e^-	1·6 & 4·3
		38	−0·028 770	321·000	0			2·9 h	e^-	1·1 & 3·0

Cl	17	32	-0·013 760	257·800	$\frac{3}{2}$			0·32 s	e^+	9·5
		33	-0·022 560	274·070	0			2·8 s	e^+	4·4
		34	-0·026 250	285·578				1·5 s	e^+	4·5
		35	-0·031 149	298·213	$\frac{3}{2}$	0·821	75·4			
		36	-0·031 691	306·790	2	1·284		3×10^5 y	e^-	0·71
		37	-0·034 101	317·106	$\frac{3}{2}$	0·683	24·6			
		38	-0·031 995	323·216	2			37 m	e^-	4·8
		39	-0·031 992	331·284				56 m	e^-	3·4
		40	-0·029 6	337·1				1·4 m	e^-	3·2 & 7·5
A	18	35	-0·024 746	291·467	$\frac{3}{2}$			1·8 s	e^+	4·9
		36	-0·032 455	306·719	0		0·34			
		37	-0·033 228	315·510	$\frac{3}{2}$	±1·000		3·5 d	EC	0·81
		38	-0·037 272	327·349	0		0·06			
		39	-0·035 683	333·940	$\frac{7}{2}$			265 y	e^-	0·565
		40	-0·037 616	343·812	0		99·6			
		41	-0·035 500	349·912	$\frac{7}{2}$			110 m	e^-	1·2 & 2·5
K	19	37	-0·026 635	308·587	$\frac{3}{2}$			1·2 s	e^+	5·1
		38	-0·030 903	320·634				7·6 m	e^+, γ	Q = 4·9
		39	-0·036 290	333·723	$\frac{3}{2}$	0·391	93·08			
		40	-0·036 000	341·524	4	-1·296	0·012	$1·3 \times 10^9$ y	{ e^- / EC	1·32 / 1·48
		41	-0·038 168	351·615	$\frac{3}{2}$	0·215	6·91	12·5 h	e^-	3·5 & 2·0
		42	-0·037 594	359·152	2	-1·137		22 h	e^-	1·8 & 0·8
		43	-0·039 270	368·784	$\frac{3}{2}$	±0·163		20 m	e^-, γ	4·9 & 1·5
		44	-0·037 960	375·640						

Element	Z	A	$(M-A)$/a.m.u.	Binding energy/ MeV	Spin	Magnetic moment	Percentage abundance	halflife	Decay mode	Decay energy/MeV
Ca	20	39	−0·029 309	326·437	$\frac{3}{2}$			1·0 s	e$^+$	6·1
		40	−0·037 411	342·056	0		96·97			
		41	−0·037 725	350·420	$\frac{7}{2}$			1·1×10⁵ y	EC	0·43
		42	−0·041 375	361·891	0	−1·595	0·64			
		43	−0·041 220	369·819	$\frac{7}{2}$		0·15			
		44	−0·044 510	380·954	0	−1·315	2·06			
		45	−0·043 811	388·374	$\frac{7}{2}$			164 d	e$^-$	0·256
		46	−0·046 311	398·775	0		0·003			
		47	−0·045 462	406·056	$\frac{7}{2}$			4·7 d	e$^-$, γ	2·0 & 0·7 1·3
		48	−0·047 469	415·996	0		0·185			
		49	−0·044 325	421·140	$\frac{3}{2}$			8·8 m	e$^-$, γ	2·0 & 1·0 3·1 & 4·1
Sc	21	41	−0·030 753	343·143	$\frac{7}{2}$			0·87 s	e$^+$	4·94
		42	−0·034 505	354·710				0·7 s	e$^+$	4·8
		43	−0·038 835	366·815	$\frac{7}{2}$	4·520		3·9 h	e$^+$	1·2
		44	−0·040 594	376·525	2	2·560		3·9 h	e$^+$, γ	1·5 1·1
		45	−0·044 081	387·844	$\frac{7}{2}$	4·749	100			
		46	−0·044 827	396·611	4	3·030		84 d	e$^-$, γ	0·36 & 1·4 0·9 & 1·1
		47	−0·047 587	407·253				3·4 d	e$^-$	0·6 & 0·4

El	Z	A			I	moment	%	half-life	decay	energy
		48	−0.047 779	415.503	6			44 h	e⁻, γ	0.64; 1.0,1.3,1.0
		49	−0.049 974	425.619				57 m	e⁻	2.0
Ti	22	44	−0.040 428	375.587	0			~10³ y	EC	0.18
		45	−0.041 871	385.003	7/2			3.1 h	e⁺	1.0
		46	−0.047 368	398.195	0		7.98			
		47	−0.048 231	407.070	5/2	−0.787	7.32			
		48	−0.052 050	418.698	0		73.99			
		49	−0.052 130	426.844	7/2	−1.102	5.46			
		50	−0.055 214	437.789	0		5.25			
		51	−0.053 397	444.168				5.8 m	e⁻	1.5 & 2.1
V	23	47	−0.045 101	403.372	7/2			33 m	e⁺	1.9
		48	−0.047 741	413.903	6			16 d	e⁺, γ	0.70; 1.0,1.3
		49	−0.051 477	425.454	7/2	±4.46		330 d	EC	0.62
		50	−0.052 836	434.791	6	3.341	0.25	4×10¹⁴ y	EC	2.4
		51	−0.056 039	445.846	7/2	5.139	99.75			
		52	−0.055 220	453.155				3.8 m	e⁻, γ	2.6; 1.4
		53	−0.056 0	462.0				2.0 m	e⁻	2.5
Cr	24	48	−0.046 240	411.720	0			23 h	EC	1.4
		49	−0.048 729	422.112	5/2			41.7 m	e⁺	1.5
		50	−0.053 946	435.042	0		4.31			
		51	−0.055 232	444.312	7/2			28 d	EC	0.75
		52	−0.059 487	456.347	0		83.76			
		53	−0.059 347	464.288	3/2	−0.473	9.55			
		54	−0.061 118	474.009	0		2.38			
		55	−0.059 167	480.263	3/2			3.6 m	e⁻	2.8

Element	Z	A	$(M-A)$/a.m.u.	Binding energy/ MeV	Spin	Magnetic moment	Percentage abundance	Decay halflife	Decay mode	Decay energy/MeV
Mn	25	51	−0·051 810	440·340				45 m	e^+	2·1
		52	−0·054 432	450·856	6	±3·08		5·7 d	e^+, γ	0·6 0·7, 0·9, 1·5
		53	−0·058 705	462·907	$\frac{7}{2}$	±5·050		2×10^6 y	EC	0·60
		54	−0·059 638	471·848	3	±3·300		290 d	EC	1·38
		55	−0·061 950	482·073	$\frac{5}{2}$	3·461	100			
		56	−0·061 090	489·343	3	3·240		2·6 h	e^-, γ	2·8, 1·0 0·8, 1·8
		57	−0·0617	498·0				1·7 m	e^-	2·6
Fe	26	52	−0·051 883	447·699	0			8·3 h	e^+	0·8
		53	−0·054 428	458·141				8·9 m	e^+	2·5
		54	−0·060 383	471·760	0		5·84			
		55	−0·061 701	481·059	$\frac{3}{2}$			2·6 y	EC	0·22
		56	−0·065 064	492·262	0		91·68			
		57	−0·064 602	499·904	$\frac{1}{2}$	<0·05	2·17			
		58	−0·066 718	509·946	0		0·31			
		59	−0·065 122	516·531	$\frac{3}{2}$			45 d	e^-	1·56, 0·46, 0·27

	Z	A	mass	spin	moment	abundance	half-life	decay	energy
Co	27	55	−0·057 987	476·817				18·2 h	EC, e+	3·46 / 1·5 & 1·0
		56	−0·060 153	486·905	4	3·86		77 d	EC, e+	4·6 / 1·5 & others
		57	−0·063 704	498·285	$\frac{7}{2}$	±4·65		270 d	EC	0·57
		58	−0·064 239	506·855	2	±4·05		71 d	EC, e+	2·3 / 0·5
		59	−0·066 811	517·321	$\frac{7}{2}$	4·639	100		e⁻,	0·3 & 1·5
		60	−0·066 187	524·848	5	±3·80		5·2 y	γ	1·2 & 1·3
		61	−0·067 560	534·162				99 m	e⁻	1·2
		62	−0·066 054	540·831				14 m	e⁻	2·9 & 0·9
Ni	28	57	−0·060 231	494·267				36 h	EC, e+	3·24 / 0·84, 0·72, 0·35
		58	−0·064 658	506·462	0		67·76			
		59	−0·065 658	515·465	$\frac{3}{2}$			10⁵ y	EC	1·075
		60	−0·069 213	526·848	0		26·16			
		61	−0·068 944	534·669	$\frac{3}{2}$	±0·750	1·25			
		62	−0·071 653	545·269	0		3·66			
		63	−0·070 336	552·108				125 y	e⁻	0·067
		64	−0·072 042	561·769	0		1·16			
		65	−0·069 928	567·872				2·6 h	e⁻	2·1, 1·0, 0·6
		66	−0·070 915	576·862	0			55 h	e⁻	0·2

Element	Z	A	$(M-A)$/a.m.u.	Binding energy/ MeV	Spin	Magnetic moment	Percentage abundance	halflife	Decay mode	Decay energy/MeV
Cu	29	58	−0.055 459	497·111				3·0 s	e⁺	8·1
		59	−0.060 504	509·882				81 d	e⁺	3·8
		60	−0.062 638	519·941	2			24 m	EC, e⁺	6·3 / 2·0, 3·0, 3·9
		61	−0.066 543	531·651	$\frac{3}{2}$	2·160		3·3 h	EC, e⁺	2·23 / 1·2
		62	−0.067 434	540·552	1			10 m	e⁺	2·9
		63	−0.070 408	551·393	$\frac{3}{2}$	2·226	69·1			
		64	−0.070 241	559·309	1	−0·216		12·8 h	EC, e⁺, e⁻	1·68 / 0·66 / 0·57
		65	−0.072 214	569·219	$\frac{3}{2}$	2·380				
		66	−0.071 129	576·279	1	±0·283		5·1 m	e⁻	2·6
		67	−0.072 241	585·386				59 h	e⁻	0·58, 0·48, 0·40

Element	Z	A			spin	moment	%	half-life	decay	energy
Zn	30	61	−0·060 75	525·47				1·5 m	e⁺ / EC	4·8 ; 1·71
		62	−0·065 620	538·079	0			9·3 h	e⁺	0·66
		63	−0·066 794	547·244	3/2			38 m	EC / e⁺	3·36 ; 2·4, 1·4
		64	−0·070 855	559·099	0		48·89			
		65	−0·070 766	567·087	5/2	0·769		245 d	EC / e⁺	1·35 ; 0·32
		66	−0·073 948	578·123	0		27·81			
		67	−0·072 855	585·175	5/2	0·873	4·11			
		68	−0·075 143	595·378	0		18·56			
		69	−0·073 459	601·881				57 m	e⁻	0·90
		70	−0·074 666	611·077	0		0·62			
		71	−0·072 490	617·120				2·2 m	e⁻	2·4
		72	−0·073 157	625·814	0			49 h	e⁻	1·6
Ga	31	64	−0·063 263	551·244				2·5 m	e⁺ / EC	6·1 & 2·8 ; 3·26
		65	−0·067 267	563·045				15 m	e⁺ / EC	2·2, 1·4, 0·8 ; 5·17
		66	−0·068 393	572·165	0			9·4 h	e⁺ / EC	4·1, 1·4, 0·9 ; 1·00
		67	−0·071 784	583·395	3/2	1·84		78 h	EC	2·9
		68	−0·072 008	591·676	1	±0·012		68 m	e⁺ / EC	1·9 & 0·8
		69	−0·074 426	602·000	3/2	2·011	60·5			
		70	−0·073 965	609·642	1			21 m	e⁻	1·65
		71	−0·075 294	618·951	3/2	2·555	39·5			
		72	−0·073 628	625·471	3	−0·132		14 h	e⁻	3·2 & others
		73	−0·074 874	634·702				5 h	e⁻	1·5
		74	−0·072 810	640·850				7·8 m	e⁻	2·6, 2·0, 1·1

Element	Z	A	$(M-A)$/a.m.u.	Binding energy/ MeV	Spin	Magnetic moment	Percentage abundance	halflife	Decay mode	Decay energy/MeV
Ge	32	67	−0·067 06	578·21				20 m	e^+	2·9
		68	−0·071 5	590·4	0			280 d	EC	0·7
		69	−0·072 037	598·992				40 h	EC ⎱ e^+	2·23 / 1·2 & 0·6
		70	−0·075 748	610·520	0		20·55			
		71	−0·075 044	617·935	$\frac{1}{2}$	0·546		11 d	EC	0·23
		72	−0·077 918	628·684	0		27·37			
		73	−0·076 537	635·470	$\frac{9}{2}$	−0·877	7·67			
		74	−0·078 819	645·667	0		36·74			
		75	−0·077 117	652·152				82 m	e^-	1·14 & others
		76	−0·078 595	661·600	0		7·67			
		77	−0·076 400	667·630				11 h	e^-	2·2, 1·4, 0·7
		78	−0·077 29	676·51	0			86 m	e^-	~1·2

	Z	A			Spin	Q	%	Half-life	Decay	Energy
As	33	70	−0·069 054	603·502				52 m	e⁺	2·5 & 1·4
		71	−0·072 887	615·144	$\frac{5}{2}$			62 h	EC; e⁺	2·01; 0·81
		72	−0·073 237	623·542	2			26 h	EC; e⁺	4·4; 3·3, 2·5,...
		73	−0·076 139	634·316	$\frac{3}{2}$			76 d	EC	0·37
		74	−0·076 067	642·321	2	1·435		18 d	EC; e⁺; e⁻	2·56; 1·5 & 0·9; 1·4 & 0·7
		75	−0·078 404	652·568	$\frac{3}{2}$		100			
		76	−0·077 603	659·894	2			26 h	e⁻	3·0, 2·4, 1·8
		77	−0·079 354	669·597	$\frac{3}{2}$			38 h	e⁻	0·68
		78	−0·078 100	676·500				91 m	e⁻	4·1
		79	−0·079 110	685·510				9·0 m	e⁻	2·3
Se	34	71	−0·068 16	609·96				5 m	e⁺	3·4
		72	−0·072 6	622·2	0			8·4 d	EC	0·5
		73	−0·073 186	630·783				7·1 h	EC; e⁺	2·75; 1·7 & 1·3
		74	−0·077 524	642·895	0		0·87			
		75	−0·077 475	650·921	$\frac{5}{2}$			121 d	EC	0·87
		76	−0·080 793	662·083	0		9·02			
		77	−0·080 089	669·498	$\frac{1}{2}$	0·532	7·58			
		78	−0·082 686	679·989	0		23·52			
		79	−0·081 506	686·961	$\frac{7}{2}$	−1·015		$<6 \times 10^4$ y	e⁻	0·16
		80	−0·083 473	696·865	0		49·82			
		81	−0·082 016	703·579	$\frac{1}{2}$			18 m	e⁻	1·4
		82	−0·083 293	712·840	0		9·19			
		83	−0·081 090	718·850	0			70 s	e⁻	3·4

Element	Z	A	$(M-A)$/a.m.u.	Binding energy/ MeV	Spin	Magnetic moment	Percentage abundance	halflife	Decay mode	Decay energy/MeV
Br	35	76	−0·075 820	656·670	1			17 h	e⁺	3·6, 1·7, …
		77	−0·078 624	667·351	$\frac{3}{2}$			58 h	EC	1·36
		78	−0·078 850	675·634	1			<6 m	e⁺	0·33
		79	−0·081 671	686·333	$\frac{3}{2}$	2·099	50·56		e⁺	2·3
		80	−0·081 464	694·212	1			18 m	EC / e⁺ / e⁻	1·89 / 0·9 / 2·0
		81	−0·083 708	704·373	$\frac{3}{2}$	2·263	49·44			
		82	−0·083 198	711·970	5	±1·6		36 h	e⁻	0·44
		83	−0·084 832	721·562				2·3 h	e⁻	0·94
		84	−0·083 450	728·350				32 m	e⁻	4·7, 3·6, 2·5
		85	−0·084 47	737·37				3 m	e⁻	2·5
Kr	36	77	−0·075 520	663·680	$\frac{5}{2}$			1·1 h	EC / e⁺	2·89 / 1·9, 1·7, 0·8
		78	−0·079 597	675·547	0		0·354			
		79	−0·079 932	683·930	$\frac{1}{2}$			34 h	EC / e⁺	1·61 / 0·6 & 0·3
		80	−0·083 620	695·437	0		2·27			
		81	−0·083 390	703·290	$\frac{7}{2}$			$2·1 \times 10^5$ y	EC	0·3
		82	−0·086 518	714·279	0		11·56			
		83	−0·085 869	721·746	$\frac{9}{2}$	−0·967	11·55			
		84	−0·088 496	732·265	0		56·90			

	A								
	85	−0·087 477	739·387	9/2	−1·001	17·37	10 y	e⁻	0·67
	86	−0·089 384	749·235	0			78 m	e⁻	3·8
	87	−0·086 635	754·745				2·8 h	e⁻	2·8
	88	−0·085 73	761·97	0			3 m	e⁻	4·0
	89	−0·083 40	767·90				35 s	e⁻	3·2
	90	−0·080 28	773·04	0			10 s	e⁻	~3·6
	91								
Rb 37	81	−0·080 980	700·270	3/2	2·05		4·7 h	{ EC / e⁺	2·26 / 1·0, 0·6, 0·3
	82	−0·082 041	709·327	1			1·2 m	e⁺	3·15
	83	−0·085 3	720·4	5/2	1·42		100 d	EC	~0·77
	84	−0·085 619	728·803	2	−1·32		33 d	{ EC / e⁺ / e⁻	2·65 / 1·6 & 0·8 / 0·91
	85	−0·088 200	739·278	5/2	1·348	72·15			
	86	−0·088 807	747·915	2	−1·69		19 d	{ EC / e⁻	0·4 / 1·8 & 0·7
	87	−0·090 813	757·855	3/2	2·741	27·85	5 × 10¹⁰ y	e⁻	0·27
	88	−0·088 73	763·99	2			18 m	e⁻	5·3, 3·6, 2·5
	89	−0·088 35	771·70				15 m	e⁻	3·9 & 2·8
	90	−0·085 18	776·82				2·7 m	e⁻	5·7

Element	Z	A	(M−A)/a.m.u.	Binding energy/ MeV	Spin	Magnetic moment	Percentage abundance	Decay halflife	Decay mode	Decay energy/MeV
Sr	38	82	−0·0816	708·1	0			25 d	EC	~0·5
		83	−0·0828 8	717·3				34 h	EC / e+	2·3 / 1·3
		84	−0·086 570	728·906	0		0·55			
		85	−0·087 011	737·388				64 d	EC	~1·1
		86	−0·090 715	748·910	0		9·87			
		87	−0·091 108	757·348	$\frac{9}{2}$	−1·089	7·02			
		88	−0·094 359	768·447	0		82·56			
		89	−0·092 558	774·840	$\frac{5}{2}$			51 d	e−	1·46
		90	−0·092 253	782·628	0			28 y	e−	0·54
		91	−0·091 839	788·45	$\frac{5}{2}$			9·7 h	e−	2·7, 1·4, ...
		92	−0·089 02	795·76	0			2·6 h	e−	1·5 & 0·55
Y	39	86	−0·085 054	742·854	4			14·6 h	e+	1·8 & 1·2
		87	−0·089 26	754·85	4			80 h	EC	1·7
		88	−0·090 472	764·044	4			105 d	EC	3·45
		89	−0·094 128	775·521	$\frac{1}{2}$	−0·137	100			
		90	−0·092 837	782·390	2	−1·630		64 h	e−	2·26
		91	−0·093 705	790·338	$\frac{1}{2}$	±0·164		58 d	e−	1·55
		92	−0·091 07	796·89	2			3·6 h	e−	3·6, 2·7, 1·3
		93	−0·090 45	804·38				11 h	e−	2·9
		94	−0·088 32	810·47				20 m	e−	5·4

	Z	A	mass excess		spin	μ	abundance (%)	half-life	decay	energy
Zr	40	87	−0·085 51	750·56				94 m	⎰EC / e⁺	3·5 / 2·1
		88	−0·0889	762·8	0			85 d	EC	2·84
		89	−0·091 086	771·905				79 h	⎰EC / e⁺	2·84 / 0·9
		90	−0·095 300	783·902	0		51·46			
		91	−0·094 358	791·096	5/2	−1·298	11·23			
		92	−0·094 969	799·736	0		17·11			
		93	−0·093 550	806·486	5/2			1·1 × 10⁶ y	e⁻	0·056
		94	−0·093 687	814·684	0		17·40			
		95	−0·091 965	821·15				65 d	e⁻	1·1, 0·9, 0·4, 0·3
		96	−0·091 714	828·99	0		2·80			
		97	−0·089 03	834·56	1/2			17 h	e⁻	1·9
Nb	41	89	−0·086 92	767·24				1·9 h	e⁺	2·9
		90	−0·089 74	777·01				15 h	e⁺	1·5 & 0·6
		91	−0·093 14	789·18				long	EC	1·2
		92	−0·092 789	796·92				10 d	EC	2·1
		93	−0·093 618	805·767	9/2	6·167	100			
		94	−0·092 70	812·98				1·8 × 10⁴ y	e⁻	0·5
		95	−0·093 168	821·49				35 d	e⁻	0·9 (1%) & 0·16
		96	−0·091 94	828·42				23 h	e⁻	0·7 & 0·4
		97	−0·092 90	836·45				72 m	e⁻	1·3

Element	Z	A	$(M-A)$/a.m.u.	Binding energy/ MeV	Spin	Magnetic moment	Percentage abundance	halflife	Decay mode	Decay energy/MeV
Mo	42	90	−0·086 06	773·73	0			6 h	EC ⎰ e⁺ ⎱	2·5 1·3
		91	−0·088 35	783·93				16 m	e⁺	3·4
		92	−0·093 190	796·514	0		15·86			
		93	−0·093 17	804·57				>2 y	EC	
		94	−0·094 910	814·26	0		9·12			
		95	−0·094 161	821·63	$\frac{5}{2}$	−0·913	15·70			
		96	−0·095 326	830·79	0		16·50			
		97	−0·093 978	837·61	$\frac{5}{2}$	−0·933	9·45			
		98	−0·094 591	846·25	0		23·75			
		99	−0·092 28	852·17				66 h	e⁻	1·2, 0·8, 0·4
		100	−0·092 525	860·47	0		9·62			
		101	−0·089 65	865·86				14 m	e⁻	2·2, 1·6, …
Tc	43	93	−0·089 75	800·60				2·7 h	EC ⎰ e⁺ ⎱	3·2 0·8
		94	−0·090 337	809·22				53 m	EC ⎰ e⁺ ⎱	4·3 2·5
		95	−0·092 38	819·19				20 h	EC	1·7
		96	−0·092 17	827·07				4·2 d	EC	2·9
		97	−0·093 7	836·5				2·6 × 10⁶ y	EC	0·3
		98	−0·092 89	843·88				1·5 × 10⁶ y	e⁻	0·3
		99	−0·093 75	852·75	$\frac{9}{2}$	5·680		2·1 × 10⁵ y	e⁻	0·29
		100	−0·092 16	859·35	1			16 s	e⁻	3·4 & 2·9

		A								
		102	−0·090 82	874·24			5 s	e⁻	4·4	
Ru	44	95	−0·090 20	816·38	0			1·7 h	e⁺	1·1
		96	−0·092 402	826·50	0		5·57	2·9 d	EC	
		97	−0·092 4	834·5						
		98	−0·094 711	844·80	0		1·86			
		99	−0·094 06	852·26	5/2	−0·630	12·7			
		100	−0·095 782	861·93	0		12·7			
		101	−0·094 423	868·74	5/2	−0·690	17·0			
		102	−0·095 652	877·96	0		31·6			
		103	−0·093 69	884·20				40 d	e⁻	0·2 & 0·7
		104	−0·094 57	893·09	0		18·5			
		105	−0·092 32	899·07				4·5 h	e⁻	1·1
		106	−0·092 68	907·47	0			1·0 y	e⁻	0·039
Rh	45	99	−0·091 81	849·38				4·5 h	{ EC / e⁺	2·1 / 0·7
		100	−0·091 87	857·51	2			20 h	{ EC / e⁺	3·6 / 2·6, 2·0, …
		101	−0·093 82	867·40				5 d	EC	
		102	−0·093 16	874·85		−0·088		210 d	{ EC / e⁺ / e⁻	2·3 / 1·2, 0·8 / 1·1
		103	−0·094 49	884·16	1/2		100			
		104	−0·093 34	891·16	1			44 s	e⁻	2·5
		105	−0·094 33	900·16				37 h	e⁻	0·57
		106	−0·092 72	906·73	1			30 s	e⁻	3·5

Element	Z	A	(M−A)/a.m.u.	Binding energy/ MeV	Spin	Magnetic moment	Percentage abundance	halflife	Decay mode	energy/MeV
Pd	46	99	−0·087 73	844·80				22 m	e⁺	2·0
		100	−0·091 23	856·13				40 d	EC	
		101	−0·091 93	864·86	0			9 h	{ EC	1·6
									e⁺ }	0·6
		102	−0·094 39	875·22	0		1·0			
		103	−0·093 89	882·82				17 d	EC	0·57
		104	−0·095 99	892·85	0		11·0			
		105	−0·094 94	899·94	$\frac{5}{2}$		22·2			
		106	−0·096 52	909·49	0		27·3			
		107	−0·094 87	916·02				7 × 10⁶ y	e⁻	0·04
		108	−0·096 11	925·25	0		26·7			
		109	−0·094 05	931·40	$\frac{5}{2}$			14 h	e⁻	1·0
		110	−0·095 84	940·20	0		11·8			
		111	−0·092 33	945·94				22 m	e⁻	2·1
		112	−0·092 61	954·28	0			21 h	e⁻	0·3
Ag	47	104	−0·091 40	887·80	5	4·000		27 m	e⁺	2·7
		105	−0·093 54	897·86	$\frac{1}{2}$	±0·101		40 d	EC	
		106	−0·093 34	905·74	1			24 m	{ e⁺	2·0
									EC }	2·98
		107	−0·094 91	915·27	$\frac{1}{2}$	−0·113	51·35			
		108	−0·094 05	922·55	1	4·200		2·3 m	{ e⁻	1·8
									EC	1·8
									e⁺	0·8

Element	Z	A			Spin	μ	Abundance	Half-life	Decay	Energy
		110	−0·093 90	938·55	1			24 s	e⁻	2·8
		111	−0·094 68	947·35	½	−0·145		7·6 d	e⁻	1·0
		112	−0·092 93	953·79	2	±0·054		3·2 h	e⁻	4·0
		113	−0·093 44	962·34	½	±0·159		5·3 h	e⁻	2·1
		114	−0·091 70	968·79				5 s	e⁻	4·6
		115	−0·091 07	976·27				21 m	e⁻	2·9
Cd	48	105	−0·090 5	894·3				55 m	e⁺	1·7
		106	−0·093 54	905·14	0		1·22			
		107	−0·093 38	913·07	5/2	−0·616		6·7 h	{EC / e⁺	1·44 / 0·32
		108	−0·095 81	923·41	0		0·88			
		109	−0·095 07	930·79	5/2	−0·829		470 d	EC	0·16
		110	−0·096 99	940·64	0		12·39			
		111	−0·095 81	947·62	½	−0·592	12·75			
		112	−0·097 24	957·02	0		24·07			
		113	−0·095 591	963·556	½	−0·619	12·26			
		114	−0·096 640	972·604	0		28·86			
		115	−0·094 57	978·75	½	−0·647		53 h	e⁻	1·1, 0·6
		116	−0·095 24	987·44	0		7·58			
		117	−0·092 76	993·20	0			50 m	e⁻	

Element	Z	A	$(M-A)$/a.m.u.	Binding energy/ MeV	Spin	Magnetic moment	Percentage abundance	halflife	Decay mode	Decay energy/MeV
In	49	109	−0·092 90	927·98	$\frac{9}{2}$	5·530		4·2 h	EC e⁺	2·0 0·8
		110	−0·092 77	935·93	2			66 m	EC e⁺	3·9 2·2
		111	−0·094 64	945·75	$\frac{9}{2}$	5·530		2·8 d	EC	0·66
		112	−0·094 46	953·64	1			15 m	e⁻ EC e⁺	2·54 1·5
		113	−0·095 91	963·07	$\frac{9}{2}$	5·523	4·33			
		114	−0·095 09	970·38	1			72 s	e⁻ EC	1·98 1·42
		115	−0·096 13	979·42	$\frac{9}{2}$	5·534	95·67	~10¹⁴ y	e⁻	0·6
		116	−0·094 68	986·14				13 s	e⁻	3·29
		117	−0·095 47	994·94	$\frac{9}{2}$			1·1 h	e⁻	0·46
Sn	50	111	−0·091 94	942·44				35 m	EC e⁺	2·5 1·5
		112	−0·095 16	953·52	0		0·95			
		113	−0·094 81	961·27				119 d	EC	0·49
		114	−0·097 23	971·59	0		0·65			
		115	−0·096 65	979·12	$\frac{1}{2}$	−0·913	0·34			
		116	−0·098 25	988·69	0		14·24			
		117	−0·097 04	995·63	$\frac{1}{2}$	−0·995	7·57			
		118	−0·098 39	1004·96	0		24·01			

	A	Δ	mass	spin	μ	abundance	$t_{1/2}$	decay	γ energies
	122	−0·096 59	1035·54	0		4·71			
	123	−0·094 26	1041·47				136 d	e⁻	1·42
	124	−0·094 73	1049·97	0		5·98			
	125	−0·092 25	1055·74				9·5 m	e⁻	2·0, 1·2
Sb 51	117	−0·095 09	993·03				2·8 h	{ EC, e⁺	2·0, 1·4, 0·7
	118	−0·094 43	1000·48				5·1 h	EC	
	119	−0·096 06	1010·08				38 h	EC	
	120	−0·094 92	1017·08	1			5·8 d	EC	0·58
	121	−0·096 18	1026·33	5/2	3·342	57·25			
	122	−0·094 82	1033·13	2	2·533		2·8 d	{ e⁻, EC	1·5
	123	−0·095 79	1042·11	7/2		42·75			
	124	−0·094 03	1048·54	3			60 d	e⁻	2·3, 1·6, ...
	125	−0·094 77	1057·30	7/2			2·0 y	e⁻	0·6, 0·3, 0·13
Te 52	120	−0·095 98	1017·28	0		0·09			
	121	−0·094 80	1024·26	0			17 d	EC	
	122	−0·096 93	1034·32	0		2·46			
	123	−0·095 72	1041·26	1/2	−0·732	0·87			
	124	−0·097 16	1050·67	0		4·61			
	125	−0·095 58	1057·27	1/2	−0·882	6·99			
	126	−0·096 67	1066·37	0		18·71			
	127	−0·094 79	1072·68				9·4 h	e⁻	0·7
	128	−0·095 52	1081·43	0		31·79			
	129	−0·093 76	1087·55				72 m	e⁻	1·4, 1·0, ...
	130	−0·093 30	1095·94	0		34·49			
	131	−0·091 42	1101·83				25 m	e⁻	2·1, 1·7, 1·4
	132	−0·091 48	1109·95	0			77 h	e⁻	0·3

Element	Z	A	(M−A)/a.m.u.	Binding energy/MeV	Spin	Magnetic moment	Percentage abundance	halflife	Decay mode	Decay energy/MeV
I	53	122	−0.092 49	1029.40				3.4 m	e+	3.1
		123	−0.094 27	1039.13	$\frac{5}{2}$			13 h	EC	
		124	−0.093 75	1046.72	2			4 d	EC	3.2
									e+	2.2, 1.5
		125	−0.095 42	1056.34	$\frac{5}{2}$	3.000			EC	0.15
		126	−0.094 37	1063.43	2			13 d	EC	2.14
									e−	1.25, 0.9, 0.4
									e+	1.1, 0.5
		127	−0.095 53	1072.59	$\frac{5}{2}$	2.808	100			
		128	−0.094 16	1079.38	1			25 m	e−	2.1, 1.7, 2.1
		129	−0.095 01	1088.25	$\frac{7}{2}$	2.617		2×10^7 y	e−	0.15
		130	−0.093 32	1094.75	5			12.5 h	e−	1.0, 0.6
		131	−0.093 87	1103.33	$\frac{7}{2}$	2.740		8.1 d	e−	0.81, 0.61, 0.33
		132	−0.092 02	1109.67	4	±3.080		2.3 h	e−	2.1, 1.5, …
		133	−0.092 25	1117.96	$\frac{7}{2}$	2.840		21 h	e−	1.3, 0.4
		134	−0.090 15	1124.07				53 m	e−	2.5, 1.5
Xe	54	124	−0.093 88	1046.05	0		0.10			
		125	−0.093 38	1053.67				18 h	EC	
		126	−0.095 71	1063.90	0		0.09			
		127	−0.094 78	1071.10				36 d	EC	
		128	−0.096 46	1080.74	0		1.92			
		129	−0.095 22	1087.66	$\frac{1}{2}$	−0.772	26.44			
		130	−0.096 49	1096.91	0		4.08			

A	Mass excess		Spin	Moment	Abundance %	Half-life	Decay	Energy
131	−0·094 21	1103·32	2		21·18			
132	−0·095 84	1112·45	0		26·89			
133	−0·094 18	1118·98	3/2			5·3 d	e⁻	0·35
134	−0·094 60	1127·44	0		10·44			
135	−0·092 98	1134·00				9·1 h	e⁻	0·9
136	−0·092 78	1141·88	0		8·87			
137	−0·088 90	1146·34				3·9 m	e⁻	3·5
138	−0·086 19	1151·89	0			17 m	e⁻	2·4
Cs 55 126	−0·090 56	1058·32	1			1·6 m	⎰ e⁺ ⎱ EC	3·8 4·8
127	−0·092 52	1068·22	1/2	±1·41		6·3 h	⎰ e⁺ ⎱ EC	0·68 2·21
128	−0·092 24	1076·03	1			3·8 m	⎰ e⁺ ⎱ EC	3·0, 2·5, 1·5 4·0
129	−0·094 04	1085·77	1/2	±1·47		31 h	EC	
130	−0·093 28	1093·14	1	1·32		30 m	⎰ e⁺ ⎨ EC ⎱ e⁻	1·97 3·00 0·44
131	−0·094 53	1103·30	5/2	3·48		10 d	EC	0·36
132	−0·093 61	1109·59	2	2·20		6·2 d	EC	
133	−0·094 64	1118·63	7/2	2·564	100			
134	−0·093 18	1125·33	4	2·973		2·1 y	e⁻	0·7, 0·3
135	−0·094 23	1134·38	7/2	2·713		3 × 10⁶ y	e⁻	0·21
136	−0·092 66	1140·99	5			13 d	e⁻	0·34, 0·66
137	−0·093 23	1149·60	7/2			27 y	e⁻	0·5
138	−0·089 20	1153·91				32 m	e⁻	3·4, 2·7
139	−0·087 10	1160·03				9·5 m	e⁻	

Element	Z	A	$(M-A)$/a.m.u.	Binding energy/ MeV	Spin	Magnetic moment	Percentage abundance	halflife	Decay mode	Decay energy/MeV
Ba	56	130	−0·093 75	1092·80	0		0·13			
		131	−0·093 28	1100·43				12 d	EC	
		132	−0·094 88	1109·99	0		0·19			
		133	−0·094 12	1117·36	$\frac{3}{2}$			7·2 y	EC	
		134	−0·095 39	1126·61	0		2·60			
		135	−0·094 45	1133·81	$\frac{3}{2}$	0·832	6·7			
		136	−0·095 70	1143·04	0		8·1			
		137	−0·094 50	1149·99	$\frac{3}{2}$	0·931	11·9			
		138	−0·095 00	1158·53	0		70·4			
		139	−0·091 40	1163·25	0			84 m	e⁻	2·2, 2·4, 0·8
		140	−0·089 43	1169·49				12·8 d	e⁻	1·0, 0·5
		141	−0·085 95	1174·32	0			18 m	e⁻	2·8
La	57	134	−0·091 34	1122·05	1			6·5 m	EC / e⁺	3·7 / 2·7
		135	−0·093 11	1131·78				19 h	EC	
		136	−0·092 62	1139·39	1			10 m	EC / e⁺	2·1
		137	−0·093 96	1148·71				6×10^4 y	EC	
		138	−0·093 09	1155·97	5	3·685	0·09	10^{11} y	EC / e⁻	1·6 / 0·2
		139	−0·093 86	1164·76	$\frac{7}{2}$	2·761	99·91			
		140	−0·090 56	1169·76	3			40 h	e⁻	2·2, 1·6, …
		141	−0·089 17	1176·53				3·8 h	e⁻	2·4

	Z	A			I			$t_{1/2}$		(MeV)
Ce	58	135	−0.090 86	1128·89				22 h	EC; e+	0·81
		136	−0.092 90	1138·88	0		0·19			
		137	−0.092 67	1146·73	0			9 h	EC	
		138	−0.094 17	1156·20	3/2		0·25	140 d	EC	0·27
		139	−0.093 57	1163·71	0		88·48			
		140	−0.094 61	1172·74	7/2	±0·89	11·07	33 d	e−	0·58
		141	−0.091 78	1178·18	0			5×10^{15} y	α	1·5
		142	−0.090 86	1185·39				33 h	e−	1·4
		143	−0.087 67	1190·50	0			285 d	e−	0·31
		144	−0.086 41	1197·39	0					
Pr	59	139	−0.091 42	1160·92				4·5 h	EC; e+; e+; EC	2·0; 1·0; 2·23; 3·25
		140	−0.090 99	1168·59	1		100			
		141	−0.092 40	1177·98	5/2	4·50				
		142	−0.090 02	1183·83	2	±0·26		19 h	e−	2·1
		143	−0.089 22	1191·16	7/2			13·7 d	e−	0·93
		144	−0.086 75	1196·93	0			17 m	e−	3·0

Element	Z	A	(M−A)/a.m.u.	Binding energy/ MeV	Spin	Magnetic moment	Percentage abundance	halflife	Decay mode	energy/MeV
Nd	60	141	−0·090 47	1175·40	$\frac{3}{2}$			2·5 h	$\begin{cases} EC \\ e^+ \end{cases}$	1·7 / 0·7
		142	−0·092 34	1185·21	0		27·3			
		143	−0·090 22	1191·31	$\frac{7}{2}$	−1·10	12·32			
		144	−0·089 96	1199·14	0		23·8	5×10^15 y	α	1·9
		145	−0·087 46	1204·88	$\frac{7}{2}$	−0·71	8·3			
		146	−0·086 91	1212·44	0		17·1			
		147	−0·083 93	1217·73	$\frac{5}{2}$	±0·59		11 d	e⁻	0·8, 0·4
		148	−0·083 13	1225·06	0		5·67			
		149	−0·079 88	1230·10	$\frac{5}{2}$			1·8 h	e⁻	1·6
		150	−0·079 08	1237·43	0		5·56			
		151	−0·076 23	1242·84				15 m	e⁻	1·9
Pm	61	145	−0·087 31	1203·95				~ 30 y	EC	0·14
		146	−0·085 37	1210·22				~ 1 y	e⁻	0·7
		147	−0·084 89	1217·85	$\frac{7}{2}$	±3·20		2·5 y	e⁻	0·22
		148	−0·082 58	1223·76	1			5·3 d	e⁻	2·5
		149	−0·081 67	1230·99	$\frac{7}{2}$			50 h	e⁻	1·0
		150	−0·079 04	1236·61				2·7 h	e⁻	2·0, 3·0
		151	−0·078 80	1244·46	$\frac{5}{2}$	±1·80		27 h	e⁻	1·1
Sm	62	143	−0·085 45	1185·30	0			9 m	e⁺	< 2·3
		144	−0·088 01	1195·75			3·0			
		145	−0·086 61	1202·52				340 d	EC	0·63

	A			I		%	$T_{1/2}$		
	147	−0·085 13	1217·29	7/2	−0·90	15·1	10^{x} y	α	2·1
	148	−0·085 21	1225·43	0		11·2			
	149	−0·082 82	1231·28	7/2	−0·75	14·0			
	150	−0·082 72	1239·26	0		7·4	93 y	e⁻	0·08
	151	−0·080 08	1244·87	7/2					
	152	−0·080 24	1253·09	0		26·8			
	153	−0·077 90	1258·98	3/2	−0·030		47 h	e⁻	0·8, 0·7, …
	154	−0·077 72	1266·88	0		22·5			
	155	−0·075 30	1272·70	3/2			23 m	e⁻	1·9
Eu 63	150	−0·080 31	1236·23	5/2			15 h	e⁻	1·1
	151	−0·080 16	1244·16	3	3·46	47·86			
	152	−0·078 25	1250·45	3	±1·92	52·14	13 y	{ EC / e⁻	1·8 / 1·8
	153	−0·078 76	1259·00	5/2	1·53		16 y	e⁻	1·8, 1·6, …
	154	−0·076 95	1265·38	3	±2·00		1·7 y	e⁻	0·15, 0·25
	155	−0·077 07	1273·57				15 d	e⁻	2·5
	156	−0·075 20	1279·90						
Gd 64	150	−0·081 39	1236·46	0			>10^{5} y	α	2·7
	151	−0·079 7	1242·9	0			150 d	EC	
	152	−0·080 20	1251·42	0		0·21			
	153	−0·078 50	1257·97	3/2			236 d	EC	0·19
	154	−0·079 07	1266·58	0		2·23			
	155	−0·077 34	1273·03	3/2	−0·27	15·1			
	156	−0·077 82	1281·56	0		20·6			
	157	−0·075 97	1287·91	3/2	−0·36	15·7			
	158	−0·075 82	1295·84	0		24·5			
	159	−0·073 63	1301·87	3/2			18 h	e⁻	0·95
	160	−0·072 88	1309·24	0		21·6			
	161	−0·070 28	1314·89	0			3·6 m	e⁻	1·9

Element	Z	A	(M−A)/a.m.u.	Binding energy/ MeV	Spin	Magnetic moment	Percentage abundance	Decay halflife	Decay mode	Decay energy/MeV
Tb	65	151	−0·076 85	1239·51				20 h	EC	
		156	−0·075 21	1278·4				5 d	EC	
		159	−0·074 65	1302·03	$\frac{3}{2}$	1·5	100	72 d	e⁻	1·7, 0·9, ...
		160	−0·072 85	1308·43	3			7 d	e⁻	0·5, 0·4
		161	−0·072 43	1316·11	$\frac{3}{2}$					
Dy	66	152	−0·075 27	1245·33	0			2·3 h	α	3·66
		153	−0·074 26	1252·46	0			5 h	α	3·78
		154	−0·075 65	1261·82	0			13 h	α	3·37
		155	−0·074 1	1268·4	0			10 h	EC	
		156	−0·076 07	1278·36	0		0·06			
		157	−0·074 7	1285·2	0			8·2 h	EC	
		158	−0·076 55	1294·02	0		0·10			
		159	−0·074 24	1300·87	0			134 d	EC	
		160	−0·074 80	1309·46	0		2·35			
		161	−0·073 05	1315·91	$\frac{5}{2}$	±0·42	19·0			
		162	−0·073 20	1324·11	0		25·5			
		163	−0·071 25	1330·37	$\frac{5}{2}$	0·58	24·9			
		164	−0·070 80	1338·02	0		28·1			
		165	−0·068 18	1343·66	$\frac{7}{2}$			139 m	e⁻	1·3, 0·9, 0·4
		166	−0·067 19	1350·81	0			82 h	e⁻	0·2

	Z	A								
		164	−0·069 61	1336·13	1	±3·3	100	37 m	EC, e⁻	1·0
		165	−0·069 58	1344·18	7/2			27 h	e⁻	1·8
		166	−0·067 71	1350·51	0			3·0 h	e⁻	1·0, 0·3
		167	−0·066 87	1357·79						
Er	68	161	−0·070 0	1311·5	0			3·1 h	EC	
		162	−0·071 26	1320·74	0		0·136	75 m	EC	
		163	−0·069 93	1327·58	0		1·56	10 h	EC	
		164	−0·070 71	1336·38						
		165	−0·069 18	1343·02	5/2		33·41			
		166	−0·069 69	1351·57	0		22·94			
		167	−0·067 94	1358·01	7/2	−0·56	27·07			
		168	−0·067 62	1365·78	0		14·88			
		169	−0·065 39	1371·78	1/2	0·513		9 d	e⁻	0·3
		170	−0·064 44	1379·96	0					
		171	−0·061 87	1384·64	5/2	±0·70		7·8 h	e⁻	1·5, 1·0
Tm	69	166	−0·066 49	1347·81	2	0·050		7·7 h	EC, e⁺	2·1
		167	−0·066 9	1356·3	1/2			9·6 d	EC	
		168	−0·065 70	1363·28				85 d	EC	
		169	−0·065 75	1371·33	1/2	−0·229	100			
		170	−0·063 94	1377·72	1	±0·240		129 d	e⁻	0·97
		171	−0·063 47	1385·35	1/2	±0·230		680 d	e⁻	0·097

Element	Z	A	$(M-A)$/a.m.u.	Binding energy/ MeV	Spin	Magnetic moment	Percentage abundance	halflife	Decay mode	Decay energy/MeV
Yb	70	168	−0·065 84	1362·56	0		0·135			
		169	−0·064 4	1369·3	$\frac{7}{2}$			32 d	EC	
		170	−0·064 98	1377·90	0		3·14			
		171	−0·063 57	1384·66	$\frac{1}{2}$	0·493	14·4			
		172	−0·063 64	1392·80	0		21·9			
		173	−0·061 94	1399·28	$\frac{5}{2}$	−0·678	16·2			
		174	−0·061 26	1406·72	0		31·6			
		175	−0·058 86	1412·55				101 h	e⁻	0·47, 0·35
		176	−0·057 32	1419·19	0		12·6			
		177	−0·054 59	1424·72	$\frac{9}{2}$			1·9 h	e⁻	1·3, 0·16
Lu	71	170	−0·061 17	1373·57				1·7 d	EC	
		171	−0·061 8	1382·3				8·1 d	EC	
		172	−0·060 7	1389·3	4			6·7 d	EC	
		173	−0·061 20	1397·81	$\frac{7}{2}$			1·4 y	EC	
		174	−0·059 65	1404·44				165 d	{EC / e⁻}	0·6
		175	−0·059 36	1412·24	$\frac{7}{2}$	2·23	97·4			
		176	−0·057 34	1418·43	7	3·18	2·6			
		177	−0·056 07	1425·32	$\frac{7}{2}$	2·24		6·8 d	e⁻	0·50, 0·38, 0·18
Hf	72	174	−0·059 64	1403·64	0		0·16			
		175	−0·058 4	1410·6				70 d	EC	
		176	−0·058 43	1418·66	0		5·21			

A	Mass excess		I	μ	Abundance	Half-life	Decay	Energy
177	$-0.056\ 60$	1425·03	$\tfrac{1}{2}$	0·61	18·56			
178	$-0.056\ 12$	1432·65	0		27·1			
179	$-0.053\ 97$	1438·72	$\tfrac{9}{2}$	-0.47	13·75			
180	$-0.053\ 18$	1446·05	0		35·22			
181	$-0.050\ 89$	1452·00				45 d	e^-	0·40
Ta 73								
176	$-0.055\ 35$	1423·08				8·0 h	EC	
177						53 h	EC	1·16
178	$-0.054\ 07$	1429·96				2·1 h	$\begin{cases} \text{EC} \\ e^+ \end{cases}$	~1
179	$-0.053\ 84$	1437·82				600 d	EC	0·094
180	$-0.052\ 46$	1444·60			0·012			
181	$-0.051\ 99$	1452·24	$\tfrac{7}{2}$	2·1	99·988			
182	$-0.049\ 83$	1458·30				115 d	e^-	0·51, 0·44, 0·36
183	$-0.048\ 53$	1654·16	$\tfrac{7}{2}$			5 d	e^-	0·6
184	$-0.046\ 02$	1470·90				8·7 h	e^-	1·26
W 74								
180	$-0.053\ 00$	1444·32	0		0·135			
181	$-0.051\ 79$	1451·27	$\tfrac{9}{2}$			145 d	EC	
182	$-0.051\ 70$	1459·26	0		26·4			
183	$-0.049\ 68$	1465·44	$\tfrac{1}{2}$	0·115	14·4			
184	$-0.048\ 97$	1472·86	0		30·6			
185	$-0.046\ 48$	1478·61	$\tfrac{3}{2}$			76 d	e^-	0·43
186	$-0.045\ 56$	1485·82	0		28·4			
187	$-0.042\ 76$	1491·28	$\tfrac{3}{2}$			24 h	e^-	1·3, 0·6

Element	Z	A	(M−A)/a.m.u.	Binding energy/MeV	Spin	Magnetic moment	Percentage abundance	Decay halflife	Decay mode	Decay energy/MeV
Re	75	181						20 h	EC	
		182	−0·048 63	1455·61				13 h	EC	
		183	−0·048 7	1463·8	$\frac{5}{2}$			64 h	EC	
		184	−0·047 2	1470·4				50 d	EC	
		185	−0·046 94	1478·26	$\frac{5}{2}$	3·172	37·07			
		186	−0·045 02	1485·50	1	±1·72		89 h	e⁻, EC	1·07
		187	−0·044 98	1492·57	$\frac{5}{2}$	3·204	62·93			
		188	−0·042 74	1498·53	1	±1·76		17 h	e⁻	2·12
Os	76	183			$\frac{9}{2}$			14 h	EC	
		184	−0·047 25	1469·69	0		0·02			
		185	−0·045 89	1476·49	$\frac{1}{2}$			94 d	EC	0·99
		186	−0·046 13	1484·79	0		1·59			
		187	−0·044 17	1491·03	$\frac{1}{2}$	0·067	1·64			
		188	−0·043 92	1498·87	0		13·3			
		189	−0·041 70	1504·88	$\frac{3}{2}$	0·657	16·1			
		190	−0·041 37	1512·64	0		26·4			
		191	−0·039 03	1518·53				16 d	e⁻	0·14
		192	−0·038 55	1526·16	0		41·0			
		193	−0·035 77	1531·64				31 h	e⁻	1·14, 1·06, …

	Z	A								
		191	−0·039 36	1518·06	$\frac{3}{2}$	0·16	38·5	74 d	$\begin{cases} e^- \\ EC \end{cases}$	2·0 0·67
		192	−0·037 30	1524·21	4					
		193	−0·036 99	1531·99	$\frac{3}{2}$	0·17	61·5	19 h	e^-	2·2, 1·9, 1·0
		194	−0·034 87	1538·10						
Pt	78	188	−0·040 33	1493·97	0			10 d	EC	
		189	−0·039 3	1501·2	0			11 h	EC	
		190	−0·040 05	1509·84	0		0·013	6×10^{11} y	α EC	3·3
		191	−0·038 5	1516·5				3·0 d	EC	
		192	−0·038 85	1524·88	0		0·78	$\sim 10^{15}$ y	α EC	2·6
		193	−0·036 94	1531·16	0					
		194	−0·037 27	1539·55	0		32·9			
		195	−0·035 19	1545·68	$\frac{1}{2}$	0·60	33·8			
		196	−0·035 03	1553·60	0		25·2			
		197	−0·032 65	1559·46	0			18 h	e^-	0·67, 0·48
		198	−0·032 10	1567·02			7·19			
		199	−0·029 42	1572·59	0			30 m	e^-	1·7, 1·3, …

Element	Z	A	$(M-A)$/a.m.u.	Binding energy/ MeV	Spin	Magnetic moment	Percentage abundance	halflife	Decay mode	Decay energy/MeV
Au	79	191	−0·0364	1513·8	$\frac{3}{2}$	±0·14		3 h	EC	
		192	−0·033 8	1518·6	1	±0·01		4·7 h	{EC, e⁺	1·9
		193	−0·0357	1529·3	$\frac{3}{2}$	±0·14		16 h	EC	
		194	−0·034 58	1536·26	1	±0·074		39 h	{EC, e⁺	2·57; 1·5, 1·2
		195	−0·034 95	1544·67	$\frac{3}{2}$	±0·148		180 d	EC	0·27
		196	−0·033 44	1551·34	2	±0·63		5·6 d	{EC, e⁻	0·30
		197	−0·033 46	1559·43	$\frac{3}{2}$	0·145	100			
		198	−0·031 77	1565·92	2	0·58		2·69 d	e⁻	0·96
		199	−0·031 23	1573·49	$\frac{3}{2}$	0·27		3·1 d	e⁻	0·46, 0·30, 0·25
		200	−0·029 30	1579·77				48 m	e⁻	2·2, 0·6
		201	−0·028 08	1586·70				26 m	e⁻	(1·5)

		A			I		%	half-life	decay	E
Hg	80	195	−0·033 4	1542·4	$\frac{1}{2}$	0·538	0·15	9·5 h	EC	
		196	−0·034 18	1551·24	0			65 h	EC	
		197	−0·032 64	1557·88	$\frac{1}{2}$	0·52				
		198	−0·033 24	1566·51	0		10·02			
		199	−0·031 72	1573·17	$\frac{1}{2}$	0·503	16·84			
		200	−0·031 67	1581·19	0		23·13			
		201	−0·029 69	1587·42	$\frac{3}{2}$	−0·607	13·22			
		202	−0·029 36	1595·18	0		28·80			
		203	−0·027 12	1601·17				47 d	e⁻	0·21
		204	−0·026 50	1608·67	0		6·85			
		205	−0·023 79	1614·21				5·5 m	e⁻	1·7
Tl	81	199	−0·030 54	1571·29	$\frac{1}{2}$	1·59		7 h	EC	
		200	−0·029 04	1577·96	2	±0·15		26 h	{EC / e⁺	2·46 / 1·4
		201	−0·029 25	1586·23	$\frac{1}{2}$	1·60		72 h	EC	0·6
		202	−0·028 05	1593·18	2	±0·15		12 d	EC	
		203	−0·027 65	1600·88	$\frac{1}{2}$	1·611	29·50			
		204	−0·026 14	1607·54	2	±0·089		3·6 y	{e⁻ / EC	0·76 / 0·38
		205	−0·025 56	1615·07	$\frac{1}{2}$	1·627	70·50			
		206	−0·023 90	1621·60				4·2 m	e⁻	1·5
		207	−0·022 55	1628·41				4·8 m	e⁻	1·45
		208	−0·017 99	1632·23	5			3·1 m	e⁻	1·8
		209	−0·014 70	1637·25				2·2 m	e⁻	1·99
(RaC″)		210	−0·009 95	1640·89				1·5 m	e⁻	1·96

Element	Z	A	$(M-A)$/a.m.u.	Binding energy/ MeV	Spin	Magnetic moment	Percentage abundance	Decay halflife	Decay mode	Decay energy/MeV
Pb	82	200	−0·028 0	1576·2	0			21 h	EC	
		201	−0·0271	1583·5				9·4 h	{ EC, e+	2·5
		202	−0·028 00	1592·35	0			3×10^5 y	EC	
		203	−0·026 77	1599·28				52 h	EC	0·95
		204	−0·026 96	1607·52	0		1·4			
		205	−0·025 52	1614·26				5×10^7 y	EC	0·11
		206	−0·025 53	1622·34	0		25·2			
		207	−0·024 10	1629·07	$\frac{1}{2}$	0·584	21·7			
		208	−0·023 35	1636·45	0		51·7			
		209	−0·018 92	1640·39				3·3 h	e−	0·63
(RaD)		210	−0·015 81	1645·57	0			19·4 y	e−	0·064
(AcB)		211	−0·011 26	1649·40				36 m	e−	1·39
(ThB)		212	−0·008 09	1654·52	0			10·6 h	e−	0·35, 0·58
(RaB)		214	−0·000 23	1663·34				26·8 m	e−	0·6, 1·0
Bi	83	203	−0·023 35	1595·31	$\frac{9}{2}$	4·59		12 h	{ EC, e+	1·35, 0·74
		204	−0·0221	1602·3	6	4·25			α, EC	4·85
		205	−0·022 62	1610·77	$\frac{9}{2}$	5·50		15 d	{ EC, e+	2·65, 0·93
		206	−0·021 61	1617·90	6	4·56		6·3 d	EC	3·7
		207	−0·021 562	1625·039				8 h	EC	2·4

	A	Mass excess	Mass (MeV)	I	μ	%	Half-life	Radiation	Energy (MeV)
	209	−0·019 606	1640·250	9/2	4·080	100			4·97
	210	−0·015 879	1644·849	1	0·044		$2·6 \times 10^{6}$ y	α	6·27, 6·18
(AcC)	211	−0·012 700	1649·960				2·16 m	e^-	2·25
(ThC)	212	−0·008 721	1654·325	1			60·5 m	α	6·08, 6·04, …
	213	−0·005 683	1659·567				46 m	e^-	1·39, 0·96
(RaC)	214	−0·001 314	1663·568				19·9 m	α	5·86
	215	+0·001 830	1668·710				8 m	e^-	1·65, 3·17, …
								α	5·5
Po 84	204	−0·019 5	1599·0	0			3·5 h	EC / α	5·37
	205	−0·018 8	1606·4	5/2	0·26			EC / α	5·3
	206	−0·019 676	1615·318	0			8·8 d	EC / α	5·22
	207	−0·018 442	1622·240	5/2	0·27		5·7 h	EC / α	5·10
	208	−0·018 757	1630·605	0			2·93 y	α	5·11
	209	−0·017 574	1637·575	1/2			103 y	α / EC	4·88 / 1·90
(RaF)	210	−0·017 124	1645·227	0			138 d	α	5·30
(AcC')	211	−0·013 343	1649·777	0			0·52 s	α	7·44
(ThC')	212	−0·011 134	1655·790	0			3×10^{-7} s	α	8·78
	213	−0·007 175	1660·174				4×10^{-6} s	α	8·35
(RaC')	214	−0·004 799	1666·032	0			$1·6 \times 10^{-4}$ s	α	7·68
(AcA)	215	−0·000 577	1670·171				$1·8 \times 10^{-3}$ s	α	7·38
(ThA)	216	+0·001 922	1675·915	0			0·158 s	α	6·78
	217	+0·006 1	1680·1				<10 s	α	6·54
(RaA)	218	+0·008 930	1685·530	0			3·05 m	α / e^-	6·00 / 0·39

Element	Z	A	$(M-A)$/a.m.u.	Binding energy/ MeV	Spin	Magnetic moment	Percentage abundance	Decay halflife	Decay mode	Decay energy/MeV
At	85	207	−0·014 440	1617·730				1·8 h	{ EC α	 5·75
		208	−0·013 4	1624·8				1·6 h	{ EC α	 5·65
		209	−0·013 833	1633·307				5·5 h	{ EC α	3·43 5·75
		210	−0·012 964	1640·569				8·3 h	{ EC α	3·93 5·63
		211	−0·012 538	1648·244				7·2 h	{ EC α	0·75 5·89
		212	−0·009 276	1653·277				0·22 s	α	~7·7
		213	−0·006 93	1659·17					α	9·2
		214	−0·003 660	1664·190				2×10^{-6} s	α	8·78
		215	−0·001 337	1670·096				10^{-4} s	α	8·00
		216	+0·002 411	1674·676				3×10^{-4} s	α	7·79
		217	+0·004 648	1680·664				0·02 s	α	7·05
		218	+0·008 607	1685·048				~2 s	α	6·63
		219	+0·011 290	1690·620				0·9 m	α	6·27

	Z	A				$T_{1/2}$	decay	energy
Rn	86	209	−0·009 6	1628·6		30 m	EC, α	6·04 / 6·04
		210	−0·010 460	1637·454	0	2·7 h	α, EC	6·04
		211	−0·009 434	1644·570		16 h	EC, α	5·85, 5·78, 5·61
		212	−0·009 293	1652·510	0	23 m	α	6·26
		215	−0·001 310	1669·290		$\sim 10^{-6}$ s	α	8·6
		216	+0·000 272	1675·886	0	$\sim 10^{-4}$ s	α	8·01
		217	+0·003 896	1680·582		$\sim 10^{-3}$ s	α	7·74
		218	+0·005 603	1687·064	0	0·019 s	α	7·13
		219	+0·009 481	1691·523		3·92 s	α	6·81, 6·55, …
		220	+0·011 401	1697·806	0	52 s	α	6·28
		221	+0·015 2	1702·3		3·82 d	e⁻, α	0·95 / 6·0
		222	+0·017 531	1708·239		3·82 d	α	5·48
Fr	87	212	−0·003 8	1646·6		19·3 m	EC, α	6·41, 6·39, 6·34
		217	+0·004 75	1679·00			α	8·3
		218	+0·007 54	1684·480		5×10^{-3} s	α	7·85
		219	+0·009 257	1690·950		0·02 s	α	7·30
		220	+0·012 337	1696·151		27·5 s	α	6·69
		221	+0·014 183	1702·504		4·8 m	α	6·33, 6·12
		222	+0·017 6	1707·4		14·8 m	e⁻, α	2·0 / 6·0
(AcK)		223	+0·019 736	1713·474		22 m	e⁻, α	1·2 / 5·3

Element	Z	A	$(M-A)$/a.m.u.	Binding energy/ MeV	Spin	Magnetic moment	Percentage abundance	halflife	Decay mode	energy/MeV
Ra	88	219	+0·010 050	1689·430				10^{-3} s	α	8·0
		220	+0·011 029	1696·587	0			0·03 s	α	7·43
		221	+0·013 892	1701·992	0			30 s	α	6·71
		222	+0·015 376	1708·681	0			37 s	α	6·55
(AcX)		223	+0·018 501	1713·842	$\frac{1}{2}$			11·6 d	α	5·87, 5·74, ...
(ThX)		224	+0·020 218	1720·314	0			3·64 d	α	5·68, 5·45
		225	+0·023 528	1725·302				14 d	e⁻	0·32
		226	+0·025 360	1731·667	0			1617 y	α	4·777
		227	+0·029 159	1736·200				41·2 m	e⁻	1·31
(MsTh₁)		228	+0·031 139	1742·427	0			6·7 y	e⁻	0·012
Ac	89	223	+0·019 144	1712·461				2·2 m	{ α EC	6·64 0·55
		224	+0·021 690	1718·160				2·9 h	{ EC α	1·37 6·17
		225	+0·023 153	1724·869				10·0 d	α	5·82, 5·78, 5·72, ...
		226	+0·026 16	1730·14				29 h	{ e⁻ EC	1·2 0·7
		227	+0·027 753	1736·727	$\frac{3}{2}$			22 y	{ e⁻ α	0·046 4·94
(MsTh₂)		228	+0·031 080	1741·700				6·13 h	e⁻	2·2, 1·8, 1·7, ...

Th	90		225	+0·023 927	1723·365			8·0 m	α, EC	6·57; 0·47
			226	+0·024 901	1730·529	0		30·9 m	α	6·33, 6·22, ...
		(RdAc)	227	+0·027 706	1735·989			18·2 d	α	6·04, 6·01, ...
		(RdTh)	228	+0·028 750	1743·087	0		1·90 y	α	5·42, 5·34, ...
			229	+0·031 652	1748·455	5/2	0·38	7340 y	α	5·02, 4·94, 4·85
			230	+0·033 087	1755·190	0		8×10⁴ y	α	4·68, 4·61
			231	+0·036 291	1760·277			26 h	e⁻	0·30, 0·22, 0·09
			232	+0·038 124	1766·741	0	100	1·4×10¹⁰ y	α	4·007
			233	+0·041 469	1771·597			22·4 m	e⁻	1·23, 1·16, ...
		(UX$_1$)	234	+0·043 583	1777·700	0		24·1 d	e⁻	0·20, 0·10
Pa	91		226	+0·027 81	1727·04			1·8 m	α	6·81
			227	+0·028 811	1734·177			38·3 m	α, EC	6·46; 0·99
			228	+0·031 010	1740·200			22 h	EC, α	2·1; 6·09, 5·85
			229	+0·032 022	1747·328	5/2		1·5 d	EC, α	0·30; 5·69
			230	+0·034 433	1753·154			17·0 d	EC, e⁻	1·24; 0·41
			231	+0·035 877	1759·881	3/2	±1·98	3·4×10⁴ y	α	5·05, 5·02, ...
			232	+0·038 612	1765·405			1·3 d	e⁻	0·3, 0·5, 1·3
			233	+0·040 132	1772·060	3/2	3·40	27 d	e⁻	0·26, 0·15, 0·57
		(UX$_2$)	234	+0·043 298	1777·182			6·7 h	e⁻	0·45, 0·32, ...
			235	+0·045 42	1783·28			23 m	e⁻	1·4

Element	Z	A	(M−A)/a.m.u.	Binding energy/ MeV	Spin	Magnetic moment	Percentage abundance	halflife	Decay mode	energy/MeV
U	92	228	+0·031 39	1739·07	0			9·3 m	α EC	6·68 0·35
		229	+0·033 48	1745·19				58 m	EC	1·36
		230	+0·033 94	1752·83	0			20·8 d	α	6·33
		231	+0·036 27	1758·73				4·3 d	α EC	5·88, 5·81, 5·66 0·37
		232	+0·037 17	1765·97	0			74 y	α α SF	5·45 5·32, 5·26, 5·13
		233	+0·039 52	1771·85	$\frac{5}{2}$	0·54		$1·6 \times 10^5$ y	α	4·82, 4·78, ...
		234	+0·040 90	1778·63	0		0·006	$2·5 \times 10^5$ y	α	4·77, 4·72
		235	+0·043 91	1783·90	$\frac{7}{2}$	±0·35	0·72	$7·1 \times 10^8$ y	α SF	4·56, 4·52, ...
		236	+0·045 64	1790·36	0			$2·4 \times 10^7$ y	α SF	4·50
		237	+0·048 61	1795·67				6·7 d	e$^-$	0·25
		238	+0·050 77	1801·73	0		99·274	$4·5 \times 10^9$ y	α SF	4·19
		239	+0·054 30	1806·51				23·5 m	e$^-$	1·2
		240	+0·056 59	1812·45	0			14·1 h	e$^-$	0·36
Np	93	234	+0·042 86	1776·03				4·4 d	EC	1·8
		235	+0·044 05	1782·99				410 d	EC α	0·13 5·06

	Z	A				I				
		236	+0·046 62	1788·66				22 h	EC / e⁻	0·92, 0·52
		237	+0·048 06	1795·40	5/2	5·00		2·2×10⁶ y	α	4·87, 4·80, ...
		238	+0·050 90	1800·83	2			2·1 d	e⁻	1·24, 0·25, ...
		239	+0·052 92	1807·91	5/2			2·3 d	e⁻	0·72, 0·65, ...
		240	+0·056 08	1812·15				7·3 m	e⁻	0·84
Pu	94	234	+0·043 31	1774·82	0			9 h	EC / α	0·43, 6·19
		235	+0·045 27	1781·07				26 m	EC / α	1·13, 5·85
		236	+0·046 07	1788·39	0			2·85 y	α	5·76, 5·72, 5·61
		237	+0·048 30	1794·39	7/2			45 d	EC / α / SF	0·23, 5·65, 5·34
		238	+0·049 51	1806·33	0			86 y	α / SF	5·49, 5·45, 5·35
		239	+0·052 15	1806·95	1/2	0·21		2·4×10⁴ y	α / SF	5·15, 5·13, ...
		240	+0·053 88	1813·42	0			6600 y	α / SF	5·16, 5·12, 5·01
		241	+0·056 74	1818·83	5/2	−0·73		13 y	e⁻ / α / SF	0·02, 4·89, 4·85
		242	+0·058 72	1825·05	0			3·8×10⁵ y	α / SF	4·90, 4·86
		243	+0·061 97	1830·09				5·0 h	e⁻	0·57, 0·49
		244	+0·064 1	1836·2	0			7×10⁷ y	α / SF	4·57
		245	+0·067 8	1840·8				10·6 h	e⁻	1·4
		246	+0·070 09	1846·75	0			10·8 d	e⁻	0·15, 0·33

Element	Z	A	(M−A)/a.m.u.	Binding energy/ MeV	Spin	Magnetic moment	Percentage abundance	halflife	Decay mode	Decay energy/MeV
Am	95	237	+0·049 84	1793·17				1·3 h	EC α	1·44 6·01
		238	+0·051 9	1798·3				1·9 h	EC	2·2
		239	+0·053 02	1805·36				12 h	EC α	0·81 5·77
		240	+0·055 3	1811·3				51 h	EC	1·3
		241	+0·056 71	1818·07	$\frac{5}{2}$	1·40		458 y	α	5·48, 5·43
		242	+0·059 50	1823·54	1	0·33		~100 y	e− EC	0·63 0·72
		243	+0·061 37	1829·88	$\frac{5}{2}$	1·40		7950 y	α	5·27, 5·22, 5·17
		244	+0·064 35	1835·16	6			26 m	e− EC	0·39 0·32
		245	+0·066 34	1841·39				2·0 h	e−	0·9
		246	+0·069 66	1846·37				25 m	e−	1·3, 1·6, 2·1

	Z	A		I	$t_{1/2}$	decay	energy
Cm	96	240	+0·055 54	1810·30	0	26·8 d	α	6·25
							SF	
		241	+0·057 54	1816·51	½	35 d	EC	0·80
							α	5·95
		242	+0·058 79	1823·42	0	163 d	α	6·11, 6·07, 5·97
							SF	
		243	+0·061 37	1829·09		35 y	α	6·06, 5·78, …
							EC	~0
		244	+0·062 82	1835·81	0	18 y	α	5·80, 5·76
							SF	
		245	+0·065 37	1841·51		8000 y	α	5·36, 5·45
		246	+0·067 20	1847·87	0	6600 y	α	5·37
							SF	
		247	+0·070 3	1853·1		4 × 10⁷ y	α	~5·2
Bk	97	243	+0·062 97	1826·82		4·5 h	EC	1·49
							α	6·72, 6·55, 6·20
		244	+0·065 2	1832·8		4·4 h	EC	2·2
							α	6·67
		245	+0·066 27	1839·88		4·98 d	EC	0·84, 6·37, 6·17, 5·89
		246	+0·068 8	1845·6		1·8 d	EC	1·46
		247	+0·070 26	1852·3		10⁴ y	α	5·51, 5·67, 5·30
		248	+0·072 96	1857·87		16 h	e⁻	0·65
							EC	0·69
		249	+0·074 883	1864·15		314 d	e⁻	0·11
							α	5·42, 5·03
							SF	
		250	+0·078 27	1869·06		3·13 h	e⁻	1·7, 0·7

Element	Z	A	(M−A)/a.m.u.	Binding energy/MeV	Spin	Magnetic moment	Percentage abundance	halflife	Decay mode	Decay energy/MeV
Cf	98	242	+0.065 31	1823·86				3·2 m	α	7·35
		243	+0.065 969	1831·31	0			12·5 m	EC, α	7·06, 7·17
		244	+0.067 905	1837·58				20 m	α	7·21, 7·17
		245	+0.068 766	1844·85	0			44 m	EC, α	1·55; 7·14
		246	+0.071 07	1850·78				36 h	α	6·75, 6·71
		247	+0.072 262	1857·74	0			2·5 h	EC	0·76
		248	+0.074 749	1863·49				350 d	α, SF	6·26
		249	+0.076 384	1870·04	0			360 y	α, SF	5·81, 5·94, …
		250	+0.079 26	1875·43				10·9 y	α, SF	6·02, 5·98
		251	+0.081 50	1881·42	0			800 y	α	6·1
		252	+0.085 02	1886·21				2·6 y	α, SF	6·11, 6·07
		253			0			17 d	e⁻	0·27
		254						56 d	SF	

	Z	A				half-life	decay	energy (MeV)
Es	99	248	+0·075 28	1854·15		25 m	α	6·87
		249	+0·076 258	1861·30		2 h	EC	1·41
		250	+0·078 61	1867·18		8 h	α	6·76
							EC	1·94
		251	+0·079 93	1874·02		1·5 d	EC	0·41
							α	6·48
		252	+0·082 81	1879·41		140 d	α	6·64
		253	+0·084 730	1885·70		20 d	α	6·63, 6·59
							SF	
		254	+0·087 90	1890·82		480 d	α	6·42
Fm	100	244				0·003 s	SF	
		245				4·2 s	α	8·15
		246				1·2 s	α	8·24
							SF	
		248	+0·077 092	1851·67	0	38 s	α	7·87, 7·83
							SF	
		249	+0·079 14	1857·84	0	2·5 m	α	7·9
		250	+0·079 49	1865·59		30 m	α	7·43
		251	+0·081 2	1872·1		7 h	EC	1·5
							α	6·89
		252	+0·082 56	1878·86	0	30 h	α	7·04
							SF	
		253	+0·084 93	1884·73		5 d	EC	0·4
							α	6·94
		254	+0·086 84	1891·02	0	3·3 h	α	7·20
							SF	
		255	+0·089 6	1896·5		21 h	α	7·1
							SF	

Element	Z	A	$(M-A)$/a.m.u.	Binding energy/ MeV	Spin	Magnetic moment	Percentage abundance	halflife	Decay mode	energy/MeV
Md	101	251	+0·084 6	1868·1				8 m	EC	
		252	+0·086 1	1874·8						
		253	+0·086 9	1882·1						
		254	+0·089 5	1887·8						
		255	+0·090 55	1894·86				30 m	EC α	0·85 7·34
		256						1·5 h	EC α	2·1 7·17
		257						3 h	EC α SF	0·5 7·07
No	102	251						0·8 s	α	8·6
		252						2·3 s	α SF	8·41
		253	+0·091 34	1877·2				105 s	α SF	8·01
		254	+0·091 14	1885·45	0			3 s	α SF	8·8
		255	+0·092 7	1892·0				15 s	α SF	8·2
		256			0			8 s	α SF	8·43
		257						23 s	α	8·23, 8·27

Rf	104	257	4·5 s	α	9·0, 8·7
		258	0·011 s	SF	
		259	3 s	α	8·77, 8·86
		260	0·3 s	SF	
		261	63 s	α	8·28
Ha	105	260	1·65 s	α	9·1

References

ABRAGAM, A. and PROCTOR, W. G. (1958), *Comptes Rendus,* vol. 246, pp. 2253–6.

ABRAGAM, A., *et al* (1962), *Physics Letters,* vol. 2, pp. 310–11.

AJZENBERG-SELOVE, F., and LAURITSEN, T. (1959), 'Energy levels of light nuclei VI', *Nuclear Physics,* vol. 11, pp. 1–340.

AJZENBERG-SELOVE, F., and LAURITSEN, T. (1966), 'Energy levels of light nuclei VII. A = 5 – 10', *Nuclear Physics,* vol. 78, pp. 1–176.

AJZENBERG-SELOVE, F., and LAURITSEN, T. (1968), 'Energy levels of light nuclei VII. A = 11 – 12', *Nuclear Physics,* vol. 114, pp. 1–142.

BETHE, H. A. (1939), *Physical Review,* vol. 55, pp. 434–56.

BHATIA, A. B., *et al.* (1952), *Philosophical Magazine,* vol. 43, pp. 485–500.

BLATT, J. M., and WEISSKOPF, V. F. (1952), *Theoretical Nuclear Physics,* Wiley.

BOHR, N., and WHEELER, J. A. (1939), *Physical Review,* vol. 56, pp. 426–50.

BREIT, G., CONDON, E. V., and PRESENT, R. D. (1963), *Physical Review,* vol. 50, pp. 825–45.

BREIT, G., *et al.* (1960), *Physical Review,* vol. 120, pp. 2227–49.

BURBRIDGE, E. M., *et al.* (1957), 'Synthesis of elements in stars', *Reviews of Modern Physics,* vol. 29, pp. 547–650.

BURGY, M. T., RINGO, G. R., and HUGHES, D. J. (1951), *Physical Review,* vol. 84, pp. 1160–4.

BUTLER, S. T. (1951), *Proceedings of the Royal Society,* series A, vol. 208, pp. 559–79.

BUTLER, S. T., and HITTMAIR, O. H. (1957), *Nuclear Stripping Reactions,* Pitman.

CHADWICK, J. (1932), *Proceedings of the Royal Society,* series A, vol. 136, pp. 692–708.

CLAUSNITZER, G. (1963), *Nuclear Instruments and Methods,* vol. 23, pp. 309–24.

COCKCROFT, J. D., and WALTON, E. T. S. (1932), *Proceedings of the Royal Society,* series A, vol. 137, pp. 229–42.

COMMINS, E. D. (1967), *Annual Review of Nuclear Science,* vol. 17, pp. 33–72.

COURANT, E. D., LIVINGSTON, M. S., and SNYDER, H. S. (1952), *Physical Review,* vol. 88, pp. 1190–6.

DANIELS, J. M. (1965), *Oriented Nuclei,* Academic Press.

DICKSON, J. M. (1965), 'Polarised ion sources and the acceleration of polarised beams', in F. J. FARLEY (ed.), *Progress in Nuclear Technology and Instrumentation,* vol. 1, pp. 103–71.

DUKE, P. J., *et al.* (1965), *Physical Review Letters,* vol. 15, pp. 468–72.

EVANS, R. D. (1955), *The Atomic Nucleus,* McGraw-Hill.

FARAGO, P. S. (1970), *Free-Electron Physics,* Penguin Books.

FARLEY, F. J. M., *et al.* (1966), *Nuovo Cimento,* vol. 45 A, pp. 281–6.

FERMI, E. (1950), *Nuclear Physics,* rev. edn, University of Chicago Press.

FERNBACH, S., HECKROTTE, W., and LEPORE, J. V. (1955), *Physical Review,* vol. 97, pp. 1059–70.

FESHBACH, H., PORTER, C. E., and WEISSKOPF, V. F. (1954), *Physical Review,* vol. 96, pp. 448–64.

FRASER, J. S., and MILTON, J. C. D. (1966), *Annual Review of Nuclear Science,* vol. 16, pp. 379–444.

FREEMANTLE, R. G., *et al.* (1953), *Physical Review,* vol. 92, pp. 1268–9.

GAMOW, G. (1938), *Physical Review,* vol. 53, pp. 595–604.

GAMOW, G., and CRITCHFIELD, C. L. (1949), *Theory of Atomic Nucleus and Nuclear Energy-Sources,* Clarendon Press.

GHIORSO, A., *et al.* (1969), *Physical Review Letters,* vol. 22, pp. 1317–20.

GHIORSO, A., *et al.* (1970), *Physical Review Letters,* vol. 24, pp. 1498–1503.

GLASSGOLD, A. E., and KELLOGG, P. J. (1958), *Physical Review,* vol. 109, pp. 1291–4.

HAHN, O., and STRASSMAN, F. (1939), *Naturwissenschaften,* vol. 27, pp. 11–5.

HEISENBERG, E. (1932), *Zeitschrift fur Physik,* vol. 77, pp. 1–11.

HEUSINKVELD, M., and FREIER, G. (1952), *Physical Review,* vol. 85, pp. 80–4.

HUGHES, I. S. (1971), *Elementary Particles,* Penguin Books.

HULL, M. H., *et al.* (1961), *Physical Review,* vol. 122, pp. 1606–19.

INGLIS, D. R. (1953), *Reviews of Modern Physics,* vol. 25, pp. 390–450.

JEFFRIES, C. D. (1960), *Physical Review,* vol. 117, pp. 1056–69.

JUVELAND, A. C., and JENTSCHKE, W. (1958), *Physical Review,* vol. 110, pp. 456–61.

KELLER, R., DICK, L., and FIDECARO, M. (1961), CERN Report 60-2, and *'Proceedings of First International Symposium on Polarisation Phenomena of Nucleons',* Basel, 1961, reprinted in *Helvetica Physica Acta,* Supplement 6, pp. 48–58.

KURTI, N., *et al.* (1956), *Nature,* vol. 178, pp. 450–3.

LATTES, C. M. G., OCCHIALINI, G. P. S., and POWELL, C. F. (1947), *Nature,* vol. 160, pp. 453–6 and 486–92.

LEIFSON, O. S., and JEFFRIES, C. D. (1961), *Physical Review,* vol. 122, pp. 1781–95.

LINHART, J. G. (1960), *Plasma Physics,* Wiley.

LIVESEY, D. L., and WILKINSON, D. H. (1948), *Proceedings of the Royal Society,* series A, vol. 195, pp. 123–34.

LIVINGSTON, M. S., and BLEWETT, J. P. (1962), *Particle Accelerators,* McGraw-Hill.

MATTAUCH, J. H. E., THIELE, W., and WAPSTRA, A. H. (1965), *Nuclear Physics,* vol. 67, pp. 1–31.

MEITNER, L., and FRISCH, O. R. (1939), *Nature,* vol. 143, pp. 239–49.

MELKONIAN, E. (1949), *Physical Review,* vol. 76, pp. 1750–9.

MENZEL, D. H. (1960), *Fundamental Formulas of Physics,* vol. 1, 2nd edn, Dover.

MOTT, N. F., and MASSEY, H. S. W. (1933), *Theory of Atomic Collisions,* Clarendon Press.

NOBLE, J. V. (1969), *Physical Review Letters,* vol. 22, pp. 473–6.

NURMIA, M., *et al.* (1967), *Physics Letters,* vol. 26B, pp. 78–80.

OCCHIALINI, G. P. S., and POWELL, C. F. (1947), *Nature,* vol. 159, pp. 186–90.

OPPENHEIMER, J. R., and PHILLIPS, M. (1935), *Physical Review,* vol. 48, pp. 500–2.

PERKINS, D. H. (1947), *Nature,* vol. 159, pp. 126–7.

PETERSON, R. E., BARSCHALL, H. H., and BOCKELMAN, C. K. (1950), *Physical Review,* vol. 79, pp. 593–7.

RAINWATER, L. J., *et al.* (1948), *Physical Review,* vol. 73, pp. 733–41.

RAMSEY, M. F. (1953), in E. Segré (ed.), *Experimental Nuclear Physics,* vol. 1, pp. 358–467, Wiley.

RASETTI, F. (1929), *Nature,* vol. 124, pp. 792–3.

REID, J. M. (1972), *The Atomic Nucleus,* Penguin Books.

RICHARDSON, J. R. (1965), 'Sector-focusing cyclotrons', F. J. FARLEY (ed.), *Progress in Nuclear Techniques and Instrumentation,* vol. 1, pp. 1–101. Wiley.

RUTHERFORD, E. (1919), *Philosophical Magazine,* vol. 37, pp. 581–7.

SALPETER, E. E. (1952), *Physical Review,* vol. 88, pp. 547–53.

SCHIFF, L. I. (1968), *Quantum Mechanics,* 3rd edn, McGraw-Hill.

SCOTT, M. J. (1958), *Physical Review,* vol. 110, pp. 1398–1405.

SEABORG, G. T. (1968), *Annual Review of Nuclear Science,* vol. 18, pp. 53–152.

SEGRÉ, E. (1964), *Nuclei and Particles,* Benjamin.

SILKELAND, T., *et al.* (1965), *Physical Review,* series B, vol. 140, pp. 277–82.

STAFFORD, G. H., *et al.* (1962), *Nuclear Instruments and Methods,* vol. 15, pp. 146–54.

STROMINGER, D., HOLLANDER, J. M., and SEABORG, G. T. (1958), 'Table of isotopes', *Reviews of Modern Physics*, vol. 30, pp. 585–904.

SUTTON, R. B., *et al.* (1947), *Physical Review*, vol. 72, pp. 1147–56.

TAYLER, R. J. (1966), *Reports on Progress in Physics*, vol. 29, pp. 489–538, reprinted in *Astrophysics*, 1969, Benjamin.

THOMPSON, W. B. (1957), *Proceedings of the Physical Society*, section B, vol. 70, pp. 1–5.

THOMPSON, W. B. (1964), *Introduction to Plasma Physics*, 2nd edn, Pergamon.

TOWNES, C. H., *et al.* (1947), *Physical Review*, vol. 71, pp. 644–5.

WAY, K., *et al.* (n.d.), *Nuclear Data Sheets*, Nuclear Research Centre, Oak Ridge, Tennessee. Appendix on nuclear moments by G. H. Fuller and V. W. Cohen.

WIGNER, E. (1937), *Physical Review*, vol. 51, pp. 106–19.

WILKINSON, D. T., and CRANE, H. R. (1963), *Physical Review*, vol. 130, pp. 852–63.

WOLFENSTEIN, L. (1949), *Physical Review*, vol. 79, pp. 1664–74.

WOLFENSTEIN, L. (1956), *Annual Review of Nuclear Science*, vol. 6, pp. 43–76.

WU, C. S., *et al.* (1957), *Physical Review*, vol. 105, pp. 1413–5.

YUKAWA, H. (1935), *Proceedings of the Physical-Mathematical Society of Japan*, vol. 17, pp. 48–57.

Index

Index